NATIONAL ACADEMIES | *Sciences Engineering Medicine*

NATIONAL ACADEMIES PRESS
Washington, DC

Companion Animals as Sentinels for Predicting Environmental Exposure Effects on Aging and Cancer Susceptibility in Humans

Ruth Cooper, Carol Berkower, and Sharyl Nass, *Rapporteurs*

National Cancer Policy Forum

Forum on Aging, Disability, and Independence

Board on Health Care Services

Health and Medicine Division

Standing Committee on the Use of Emerging Science for Environmental Health Decisions

Division on Earth and Life Studies

Proceedings of a Workshop

THE NATIONAL ACADEMIES PRESS 500 Fifth Street, NW Washington, DC 20001

This activity was supported by the American Kennel Club Canine Health Foundation; Animal Cancer Foundation; College of Veterinary Medicine & Biomedical Science at Texas A&M University; Environmental Protection Agency, Contract No. 68HERC19D0011 (Task Order No. 68HERC22F0077); Flint Animal Cancer Center, Colorado State University; Morris Animal Foundation; National Cancer Institute, Contract No. HHSN263201800029I (Task Order No. HHSN26300008); National Institute of Environmental Health Sciences, Contract No. HHSN263201800029I (Task Order No. 75N98020F0018); National Institute on Aging/National Institutes of Health, Contract No. HHSN2632018000029I (Task Order 75N98019F00848); Nicholas School of the Environment; North Carolina State University College of Veterinary Medicine; and University of Colorado Cancer Center.

International Standard Book Number-13: 978-0-309-68794-2
International Standard Book Number-10: 0-309-68794-2
Digital Object Identifier: https://doi.org/10.17226/26547

This publication is available from the National Academies Press, 500 Fifth Street, NW, Keck 360, Washington, DC 20001; (800) 624-6242 or (202) 334-3313; http://www.nap.edu.

Copyright 2022 by the National Academy of Sciences. National Academies of Sciences, Engineering, and Medicine and National Academies Press and the graphical logos for each are all trademarks of the National Academy of Sciences. All rights reserved.

Printed in the United States of America.

Suggested citation: National Academies of Sciences, Engineering, and Medicine. 2022. *Companion animals as sentinels for predicting environmental exposure effects on aging and cancer susceptibility in humans: Proceedings of a workshop*. Washington, DC: The National Academies Press. https://doi.org/10.17226/26547.

The **National Academy of Sciences** was established in 1863 by an Act of Congress, signed by President Lincoln, as a private, nongovernmental institution to advise the nation on issues related to science and technology. Members are elected by their peers for outstanding contributions to research. Dr. Marcia McNutt is president.

The **National Academy of Engineering** was established in 1964 under the charter of the National Academy of Sciences to bring the practices of engineering to advising the nation. Members are elected by their peers for extraordinary contributions to engineering. Dr. John L. Anderson is president.

The **National Academy of Medicine** (formerly the Institute of Medicine) was established in 1970 under the charter of the National Academy of Sciences to advise the nation on medical and health issues. Members are elected by their peers for distinguished contributions to medicine and health. Dr. Victor J. Dzau is president.

The three Academies work together as the **National Academies of Sciences, Engineering, and Medicine** to provide independent, objective analysis and advice to the nation and conduct other activities to solve complex problems and inform public policy decisions. The National Academies also encourage education and research, recognize outstanding contributions to knowledge, and increase public understanding in matters of science, engineering, and medicine.

Learn more about the National Academies of Sciences, Engineering, and Medicine at **www.nationalacademies.org**.

Consensus Study Reports published by the National Academies of Sciences, Engineering, and Medicine document the evidence-based consensus on the study's statement of task by an authoring committee of experts. Reports typically include findings, conclusions, and recommendations based on information gathered by the committee and the committee's deliberations. Each report has been subjected to a rigorous and independent peer-review process and it represents the position of the National Academies on the statement of task.

Proceedings published by the National Academies of Sciences, Engineering, and Medicine chronicle the presentations and discussions at a workshop, symposium, or other event convened by the National Academies. The statements and opinions contained in proceedings are those of the participants and are not endorsed by other participants, the planning committee, or the National Academies.

Rapid Expert Consultations published by the National Academies of Sciences, Engineering, and Medicine are authored by subject-matter experts on narrowly focused topics that can be supported by a body of evidence. The discussions contained in rapid expert consultations are considered those of the authors and do not contain policy recommendations. Rapid expert consultations are reviewed by the institution before release.

For information about other products and activities of the National Academies, please visit www.nationalacademies.org/about/whatwedo.

PLANNING COMMITTEE ON THE ROLE OF COMPANION ANIMALS AS SENTINELS FOR PREDICTING ENVIRONMENTAL EXPOSURE EFFECTS ON AGING AND CANCER SUSCEPTIBILITY IN HUMANS[1]

LINDA S. BIRNBAUM (*Chair*), Scientist Emeritus, National Institute of Environmental Health Sciences/National Toxicology Program; Scholar in Residence, Nicholas School of the Environment, Duke University

MATTHEW BREEN, Professor of Genomics, Oscar J. Fletcher Distinguished Professor of Comparative Oncology Genetics, Department of Molecular Biomedical Sciences, College of Veterinary Medicine, North Carolina State University

MYRTLE DAVIS, Executive Director, Discovery Toxicology, Bristol Myers Squibb

NICOLE DEZIEL, Associate Professor, Yale School of Public Health, Yale Center for Perinatal, Pediatric, and Environmental Epidemiology

WILLIAM FARLAND, Professor Emeritus, Environmental and Radiological Health Sciences, College of Veterinary Medicine and Biomedical Sciences, Colorado State University

ROY JENSEN, Director, The University of Kansas Cancer Center, Kansas Masonic Cancer Research Institute; William R. Jewell, MD Distinguished Masonic Professor, The University of Kansas Cancer Center, The University of Kansas Medical Center

DANIEL PROMISLOW, Principal Investigator and Co-Director, Dog Aging Project; Professor, Department of Lab Medicine & Pathology and Department of Biology, University of Washington School of Medicine

WENDY SHELTON, Principal, Virtual Beast Consulting; Consultant, Colorado State University

CHERYL LYN WALKER, Alkek Presidential Chair in Environmental Health; Director, Center for Precision Environmental Health; Professor, Departments of Molecular & Cell Biology and Medicine, Baylor College of Medicine

Project Staff

RUTH COOPER, Associate Program Officer
TOCHI OGBU-MBADIUGHA, Senior Program Assistant

---

[1] The National Academies of Sciences, Engineering, and Medicine's planning committees are solely responsible for organizing the workshop, identifying topics, and choosing speakers. The responsibility for the published Proceedings of a Workshop rests with the workshop rapporteurs and the institution.

MARILEE SHELTON DAVENPORT, Senior Program Officer, Standing Committee on the Use of Emerging Science for Environmental Health Decisions (until March 2022)
TRACY A. LUSTIG, Senior Program Officer; Director, Forum on Aging, Disability, and Independence
SHARYL NASS, Senior Director, Board on Health Care Services; Co-Director, National Cancer Policy Forum

Reviewers

This Proceedings of a Workshop was reviewed in draft form by individuals chosen for their diverse perspectives and technical expertise. The purpose of this independent review is to provide candid and critical comments that will assist the National Academies of Sciences, Engineering, and Medicine in making each published proceedings as sound as possible and to ensure that it meets the institutional standards for quality, objectivity, evidence, and responsiveness to the charge. The review comments and draft manuscript remain confidential to protect the integrity of the process.

We thank the following individuals for their review of this proceedings:

BERNADETTE DUNHAM, George Washington University
BEVERLY H. KOLLER, University of North Carolina at Chapel Hill
LAUREN TREPANIER, University of Wisconsin–Madison

Although the reviewers listed above provided many constructive comments and suggestions, they were not asked to endorse the content of the proceedings nor did they see the final draft before its release. The review of this proceedings was overseen by **ELI Y. ADASHI,** Brown University. He was responsible for making certain that an independent examination of this proceedings was carried out in accordance with standards of the National Academies and that all review comments were carefully considered. Responsibility for the final content rests entirely with the rapporteurs and the National Academies.

We also thank staff member Alexandra Beatty for reading and providing helpful comments on this manuscript.

Acknowledgments

The National Academies of Sciences, Engineering, and Medicine's Board on Health Care Services wishes to express its sincere gratitude to the planning committee chair, Linda S. Birnbaum, for her valuable contributions to the development and orchestration of this workshop. The board also wishes to thank all the members of the planning committee, who collaborated to ensure a workshop replete with informative presentations and moderated rich discussions. We are also grateful for the support of our workshop sponsors, without which we could not have undertaken this project, particularly Danielle Carlin, National Institute of Environmental Health Sciences, and Rodney Page, Flint Animal Cancer Center, Colorado State University. Finally, the board wants to thank the speakers, who generously shared their expertise and their time with workshop participants. Research assistance was provided by Christopher Lao-Scott, National Academies.

Contents

ACRONYMS AND ABBREVIATIONS	xvii
PROCEEDINGS OF A WORKSHOP	1
INTRODUCTION	1
BACKGROUND ON CANCER, AGING, AND ENVIRONMENTAL EXPOSURE RESEARCH	6
HISTORY AND CURRENT STATE OF THE SCIENCE OF ENVIRONMENTAL EXPOSURE EFFECTS ON AGING AND CANCER SUSCEPTIBILITY	10

 Environmental Exposure and Cancer, 10
 Environmental Exposure, Cancer, and Aging in Companion Animals: The Companion Dog Model, 15
 Why Pet Dogs with Spontaneous Tumors are Good Models for Human Disease, 18
 How Diet Modulates the Tumor Microenvironment (TME), 23
 Epigenetic Aging as a Target and Biomarker for Environmental Exposures, 25
 Domestic Dogs as a System for Understanding Aging and Life Span, 27
 Aging, Somatic Evolution, and Cancer—The Inexorable Link, 29
 Discussion: Animal Size, Reproductive Status, and Cancer, 31

METHODS AND CURRENT STUDIES	33

 The Exposome and Health, 33

Biomonitoring of Chemical Exposure in Companion Animals
  and Humans, 37
Ongoing Canine Population Studies, 40
Discussion: Advancing Use of Companion Dogs as Sentinels, 51
RELEVANCE OF COMPANION ANIMAL EXPOSURES
TO HUMAN CANCER AND AGING   54
  Exposures to Air Pollution, Smoking, and Lead, 54
  Using Silicone Samplers to Assess Air Pollution in
    People and Pets, 56
  Cats as Sentinels for Persistent Organic Pollutants Indoors, 58
  Radon Exposures and Cancer in Pets, 61
  Heavy Metal Exposures in the Dogs of Chernobyl, 62
  Exposures through Food, 64
  Chemical Mixtures, Cancer, and Diet—The Example
    of Pesticides, 65
  Discussion: Relevance of Companion Animal Exposures
    to Humans, 68
ACCELERATING CROSS-SPECIES COMPARISONS:
OPPORTUNITIES AND CHALLENGES IN DATA
SOURCES, COLLECTION, STORAGE, MODELING,
AND SHARING   71
  Human Exposure Assessment, 71
  Comparative Oncology: How Dogs are Helping Researchers
    Understand and Treat Cancer, 73
  Biobanks for Companion Animal Sentinel Studies, 76
  Using Ontologies to Unify Genomics and Phenomics
    across Species, 79
  Data, Samples, and Modeling, 81
  Discussion: Accelerating Cross-Species Comparisons, 85
EQUITY, ETHICS, AND POLICY   90
  Ethical Considerations of Using Companion Animals as
    Sentinels: Research Subject Protections, Citizen
    Science Issues, and Shared Health, 90
  One Health Approaches in Arctic Indigenous Communities, 93
  Aligning Health Care for a Bonded Family Society, 95
  Discussion: Equity, Ethics, and Policy, 96
IDENTIFYING RESEARCH GAPS AND SETTING A
RESEARCH AGENDA: EXPLORING NEXT STEPS FOR
THE PATH FORWARD   99

Obtaining Data on Pets and Exposures, 99
   Interdisciplinary Training of Researchers, Physicians, and
      Veterinarians for One Health, 100
   Translating the Science between Pets and People, 101
   Data Integration, 102
   Data Collection: Long-Term Investment, 103
   Data Collection: Tapping into Existing Cohorts (Short Term), 103
   Continuing and Expanding the Conversation, 104
   Community Research—Engagement and Equity, 104
CONCLUDING REMARKS AND POTENTIAL
NEXT STEPS   105
   References, 107

APPENDIXES
A   STATEMENT OF TASK   125
B   WORKSHOP AGENDA   127
C   BIOGRAPHICAL SKETCHES OF PLANNING
    COMMITTEE MEMBERS AND
    WORKSHOP SPEAKERS   133

Boxes, Figures, and Table

BOXES

1 Suggestions from Individual Workshop Participants to Advance the Use of Companion Animals as Sentinels for Predicting Environmental Exposure Effects on Aging and Cancer Susceptibility in Humans, 3
2 Examples of NIEHS-Sponsored Sensor Research, 41
3 The Golden Retriever Lifetime Study, 44
4 The Dog Aging Project, 46
5 Studies of Dogs as Sentinels for Testicular Dysgenesis Syndrome, 49

FIGURES

1 Increasing incidence of common cancers and specifically colorectal cancers (CRC): Environmental etiology?, 11
2 The complexities of documenting exposures, 13
3 Dogs can serve as a useful model in comparative oncology, 17
4 The remarkable similarity in cancers that develop in humans and dogs, 18
5 The nonlinear relationship between dog age and human age, 28
6 Cancers requiring different numbers of driver mutations and originating from stem cell pools that are organized in vastly different ways demonstrate very similar age-dependent incidence, 30
7 The exposome: Implications for examining environmental health disparities, 34

8   The exposome concept, 35
9   Approaches to exposure science at the NIEHS, 40
10 Contingency Test for Trend based on (A) associations of body weight and (B) body condition scores with quartile rankings based on serum total PFAS concentrations, 60
11 The Hallmark Framework of cancer progression, 66
12 Comparative molecular features of canine and human osteosarcomas, 75
13 Development of species-agnostic ontologies to classify phenotypes across species, 80
14 Integration of exposure event modeling with the Monarch Knowledge Graph, 82

TABLE

1   A Selection of Canine Cancer Sentinel Studies Showing Positive Association or No Association with Environmental Exposure, 21

Acronyms and Abbreviations

AAVSB	American Association of Veterinary State Boards
ACVO	American College of Veterinary Ophthalmologists
AD	Alzheimer's disease
AI	artificial intelligence
AIDS	acquired immunodeficiency syndrome
AKC	American Kennel Club
AML	acute myeloid leukemia
ATSDR	Agency for Toxic Substances and Disease Registry
AVCC	Access to Veterinary Care Coalition
AVMA	American Veterinary Medical Association
BMI	body mass index
CA	conformity assessment
C-BARQ	Canine Behavioral Assessment & Research Questionnaire
CBPR	community-based participatory research
CDC	Centers for Disease Control and Prevention
CE	continuing education
CHIC	Canine Health Information Center
CKD	chronic kidney disease
CLL	chronic lymphocytic leukemia
CML	chronic myeloid leukemia
CO	carbon monoxide
COHA	Clinical and Translational Science Award One Health Alliance

COTC	(NCI) Comparative Oncology Trials Consortium
CRC	colorectal cancer
CRDC	Cancer Research Data Commons
CREID	Centers for Research in Emerging Infectious Diseases
CTL	cytotoxic T lymphocyte
CTSA	clinical and translational science awards
DAP	Dog Aging Project
DDT	dichloro-diphenyl-trichloroethane
DEHP	di(2-ethylhexyl)phthalate
DEMS	Diesel Exhaust in Miners Study
DOHAD	Developmental Origins of Health and Disease
EMR	electronic medical record
EMT	epithelial mesenchymal transition
EPA	Environmental Protection Agency
ER+	estrogen-receptor-positive
EWAS	exposome-wide association study
FDA	Food and Drug Administration
FH	feline hyperthyroidism
fMRI	functional magnetic resonance imaging
GC	gas chromatography
GIS	geographic information systems
GO	Gene Ontology
GRLS	Golden Retriever Lifetime Study
GXE	gene by environment
HHEAR	Human Health Exposure Analysis Resource
HRMS	high-resolution mass spectrometry
IACUC	Institutional Animal Care and Use Committee
IARC	World Health Organization International Agency for Research on Cancer
ICDC	Integrated Canine Data Commons
ICP-MS	inductively coupled plasma mass spectroscopy
IGF	insulin-like growth factor
ISBER	International Society for Biological and Environmental Repositories
MESA	Multi-Ethnic Study of Atherosclerosis

MET	mesenchymal to epithelial transition
MHC	major histocompatibility complex
ML	machine learning
MOU	memorandum of understanding
MS	mass spectrometry
MWAS	metabolome-wide association study
NASEM	National Academies of Sciences, Engineering, and Medicine
NCI	National Cancer Institute
NGO	nongovernmental organization
NHANES	National Health and Nutrition Examination Survey
NIA	National Institute on Aging
NIEHS	National Institute of Environmental Health Sciences
NIH	National Institutes of Health
NIST	National Institutes of Standards and Technology
NSRL	no-significant-risk level
NTP	National Toxicology Program
OFA	Orthopedic Foundation for Animals
PATO	Phenotype and Trait Ontology
PBDE	polybrominated diphenyl ether
PBMC	peripheral blood mononuclear cell
PBPK	physiologically based pharmacokinetics
PCB	polychlorinated biphenyl
PDMS	polydimethylsiloxane
PEGS	Personalized Environment and Genes Study (NIEHS)
PET/CT	positron emission tomography–computed tomography
PFAS	poly- and perfluoroalkyl substances
PFHxS	perfluorohexane sulfonate
PFOA	perfluorooctanoic acid
PFOS	perfluorooctane sulfonic acid
PHD3	prolyl-hydroxylase 3
PI	principal investigator
PK	pharmacokinetics
POC	point of care device
POP	persistent organic pollutant
PPN	primary pulmonary neoplasia
RACE	Registry of Approved Continuing Education

SARS	severe acute respiratory syndrome
SES	socioeconomic status
SHS	secondhand smoke
SNP	single nucleotide polymorphism
STR	short tandem repeat
TBT	tributyltin
TDCIPP	tris(1,3-dichloroisopropyl)phosphate
TDS	testicular dysgenesis syndrome
TME	tumor microenvironment
TRI	Toxics Release Inventory program
TRIAD	Test of Rapamycin in Aging Dogs clinical trial
TVMDL	Texas A&M Veterinary Medical Diagnostic Laboratory
UFP	ultrafine particle
VOC	volatile organic compound
WHICAP	Washington Heights/Inwood Columbia Aging Project
WHO	World Health Organization

Proceedings of a Workshop

INTRODUCTION

Pet dogs and cats breathe the same air, sit on the same couch, play in the same yard, drink the same water, and often sleep in the same bed as their human companions. Because they share the environment of humans, these companion animals are exposed to the same environmental agents. Cats and dogs are also susceptible to many of the same age-related diseases as humans, including cancers associated with environmental risk factors, although they age more rapidly and show the effects of exposures sooner. Given the lifelong immersion of companion animals in the natural human environment, data collected on their exposures and health outcomes have the potential to provide new insights into environmentally induced disease that complement traditional laboratory, clinical, and public health studies. Furthermore, because companion animals experience a similar spectrum of disease as humans, with earlier onset, they could serve as sentinels for environmental risks to human health.

To examine the potential role of companion animals as sentinels of relevant, shared environmental exposures that may affect human aging and cancer, the National Cancer Policy Forum held a workshop[1] in collaboration

---

[1] The planning committee's role was limited to planning the workshop, and the Proceedings of a Workshop has been prepared by the workshop rapporteurs as a factual summary of what occurred at the workshop. Statements, recommendations, and opinions expressed are those of individual presenters and participants, and are not necessarily endorsed or verified by the National Academies of Sciences, Engineering, and Medicine, and they should not be construed as reflecting any group consensus.

1

with the Forum on Aging, Disability, and Independence and the Standing Committee on the Use of Emerging Science for Environmental Health Decisions to explore this promising and underutilized pathway for research. This hybrid workshop was held at the Keck Center in Washington, DC, and concurrently online on December 1–3, 2021. The planning committee developed the agenda for the workshop sessions, selected and invited speakers and discussants, and moderated the panel discussions. In designing the workshop, the planning committee focused on identifying gaps in the research and solutions to close them.

The workshop convened an array of experts in diverse fields representing academic veterinary and medical centers, universities and schools of public health, government agencies, industry, and pet owners. The expertise of the invited speakers included environmental exposures, comparative oncology, biomonitoring, data management, veterinary bioethics, and environmental justice, as well as other areas. The workshop was open to the public and audience members also included a variety of perspectives. Presentations and panel discussions covered the current state of the science and pathways for accelerating research, along with opportunities and challenges for using this novel translational approach to exposure science to advance human health. This workshop proceedings is the rapporteurs' summary of the speakers' presentations and the moderated panel discussions. The moderated discussions included panelists' responses to questions from both the in-person and virtual audience, as well as audience comments; when identified, audience comments are attributed by name and affiliation. The workshop statement of task is in Appendix A, the agenda in Appendix B, and the biosketches of workshop planning committee members and invited speakers in Appendix C. The webcast and speakers' presentations have been archived online.[2]

Planning committee chair Linda S. Birnbaum (National Institute of Environmental Health Sciences [NIEHS] and National Toxicology Program [NTP], Emeritus; Duke University) opened the workshop by outlining the topics to be explored, which included:

- Potential data sources needed to assess whether companion animals may serve as sentinels for human environmental exposures;
- The state of the science for biomarkers of exposure and use of biosensors for application to companion animal populations of interest;
- Best practices for collection, storage, and analysis of biosamples to assess exposures;

---

[2] See https://www.nationalacademies.org/event/12-01-2021/the-role-of-companion-animals-as-sentinels-for-predicting-environmental-exposure-effects-on-aging-and-cancer-susceptibility-in-humans-a-workshop (accessed December 10, 2021).

- Strategies for standardizing, sharing, and aggregating health records and relevant metadata across species; and
- Current policies and regulations related to monitoring and mitigating environmental exposures and the role for prospective interventions based on companion animal data.

Over the course of 3 days, participants offered many suggestions to advance the use of companion animals as sentinels for human environmental exposures, cancer, and aging, and these are summarized in Box 1. Additional details regarding these suggestions can be found throughout the proceedings.

BOX 1
Suggestions from Individual Workshop Participants to Advance the Use of Companion Animals as Sentinels for Predicting Environmental Exposure Effects on Aging and Cancer Susceptibility in Humans[a]

Collecting, standardizing, and analyzing data
- Develop enhanced infrastructure for collecting and standardizing demographic and genomic information from observational cohort studies. (Page)
- Upload data from all companion dog studies to the National Cancer Institute's (NCI's) Integrated Canine Data Commons (ICDC). (LeBlanc, Promislow, Sharpless)
- Expand the ICDC to include exposure data and other types of data across species. (Page)
- Standardize veterinary electronic medical records. (Breen, Deziel, Thessen)
- Standardize and share all the canine '-omic data that are currently being accumulated. (Ostrander, Walker)
- Develop a registry for longitudinal dog studies, through the American Veterinary Medical Association (AVMA) clinical trials database or the American Kennel Club (AKC). (Hughes, LeBlanc)
- Develop dog and cat phenotype ontologies that can be integrated into multispecies knowledge graphs. (Thessen)
- Develop a repository of data on the composition of commercial dog foods, including heavy metal and heterocyclic amine contents. (Wakshlag)
- Use the standardized instrumentation and protocols of the Human Health Exposure Analysis Resource (HHEAR) network

continued

BOX 1 CONTINUED

to analyze large exposome data sets from canine studies, particularly for untargeted searches where standards do not exist. (Miller)
- Explore opportunities to automate discovery through large data sets. (Birnbaum, Boyko)

Establishing collaborations between veterinary and human research and clinical practice
- Develop companion animal studies ancillary to existing human studies, particularly ones that are recruiting new cohorts, such as NCI-CONNECT. (Jones)
- Include animals when building longitudinal cohorts for observational studies of potential hazards, such as those related to natural gas drilling. (Rabinowitz)
- Develop studies that allow physicians and veterinarians to collect biospecimens from humans and pets simultaneously—for example, through existing mobile clinics. (Carlin)
- Convene leaders of human and pet prospective studies to identify potential collaborations. (Promislow)
- Enroll human pet owners in longitudinal pet studies. (Birnbaum, Promislow)
- Provide supplemental NIEHS/NCI/National Institute on Aging funding to add pet biomonitoring and outcome data to existing longitudinal studies. (Birnbaum)
- Establish a mechanism for a veterinarian who detects an environmental hazard in a companion animal to convey that information to clinicians working with the human household. (Rabinowitz)
- Create positions for veterinarians on local public health boards. (Ruple)

Addressing the denominator problem to obtain more accurate data on U.S. dogs
- Assemble a team of human demographers, geospatial scientists, and industry representatives to develop models to estimate denominators based on breed, age, size, and so on, using existing databases (Promislow)
- Systematically include the reason for euthanasia in dogs' veterinary records. (Ruple)
- Work with the AKC, AVMA, researchers, and industry to restore the question about pets in the household to the 2030 U.S. Census. (Dunn, Jones, LeBlanc)

Building capacity in veterinary schools and clinics
- Encourage veterinary, medical, and public health students to train in One Health, including offering public health courses dual DVM-MPH degrees in veterinary schools. (Birnbaum, Johannes, Ryan, Wakshlag)
- Support teaching of comparative medicine in veterinary and medical schools. (Carlin)
- Establish a board certification system for veterinary geriatric specialists. (Promislow)
- Establish a national program for schools of veterinary medicine to facilitate the study of spontaneous cancers in companion animals as potential models for human cancer. (Jensen)
- Provide continuing education (CE) opportunities and credits to veterinarians participating in research. (Promislow, Ruple)
- Build community research education into veterinary conferences and CE expectations for licensure. (Page)

Designing prospective studies to account for population variation, diversity, and complexity
- Investigate exposures in diverse populations. (Ellison, Ruple)
- Gather more data on cancer in disadvantaged populations that are currently underrepresented in studies. (Jones)
- Consider variations in genetic susceptibility and windows of exposure in prospective study design. (Ellison)
- Expand cancer research in aging animals. (Haigis)
- Explore how to better incorporate working dogs into the One Health framework. (von Hippel)

Moving the science forward on exposures
- Assemble working groups to identify important exposures for further study. (Breen, Miller, Promislow, Wakshlag)
- Convene an expert panel to evaluate the risks of pesticide exposures. (Wakshlag)
- Convene a panel to evaluate the potential long-term health risks of heterocyclic amine and acrylamide exposure in dogs and humans and reconsider safe upper limits for these compounds. (Wakshlag)
- Leverage existing environmental data and exposure information collected by the NIEHS, the U.S. Environmental Protection Agency, and state and local health departments for estimating exposures. (Farland, Jones)
- Evaluate tissue-specific reactions to exposures. (Ellison)
- Develop methods to identify "selectogens"—agents that increase the cancer risk by altering cellular context rather than inducing mutations. (DeGregori)

continued

> **BOX 1 CONTINUED**
>
> Develop bioassays to study selectogens and agents that cause chronic inflammation. (Birnbaum, Sharpless)
> - Evaluate cancer risks associated with complex mixtures and low doses of agents that affect large populations. (Ellison, Jones, Ryan)
> - Incorporate biologic and mechanistic components in prospective designs. (Ellison)
> - Engage bioinformaticians, genetic epidemiologists, and statistical geneticists in planning canine exposure studies, to obtain results that meet human research standards of statistical validity. (Ostrander)
>
> **Advancing community engagement and equity in research**
> - Practice community-engaged research, ideally in the form of community-based participatory research. (von Hippel)
> - Engage a research ethicist and an educator in the planning phase. (Moses)
> - Include learning objectives to enhance participants' science literacy and engage local teachers. (Moses)
> - Seek additional review of research plans by a Veterinary Clinical Trial Ethical Review Committee. (Moses)
> - Collaborate with social workers to enhance communication and shared decision making with participants during enrollment (Moses)
> - Avoid coercive enrollment practices—for example, by offering veterinary clinics to community members regardless of participation. (Moses)
> - Seek to gain cultural competency and partner with organizations that have existing relationships in the community and with local leaders. (Moses)
> - Establish memoranda of understanding (MOUs) between research organizations and communities that clearly state relationship, data sovereignty, and other issues important for community participation. (von Hippel)

BACKGROUND ON CANCER, AGING, AND ENVIRONMENTAL EXPOSURE RESEARCH

Ned Sharpless (National Cancer Institute [NCI]) placed this workshop into context with respect to the current state of cancer research and environmental exposures, as well as research needs. There have been significant advances in cancer treatment in recent years, most dramati-

- Communicate research progress and results back to communities, using accessible channels, and enable community members to contact researchers. (von Hippel)
- Return data derived from individual participants back to those individuals. (Birnbaum, Carlin)

Standardizing, supporting, and using biobanks
- Standardize biobanking protocols and consent procedures. (Castelhano)
- Establish a national, centralized environmental specimen biobanking initiative for human and sentinel species. (Castelhano)
- Develop marketing/communication initiatives to promote awareness of biobank resources. (Castelhano)
- When developing biobanks, aim for compliance with international standards and conformity assessment through accreditation. (Castelhano)
- Make the large number of existing biobank specimens available for study. (Farland)
- Provide biospecimen science funding to support long-term maintenance of biobank specimens. (Castelhano)

Growing support within and outside the research community
- Disseminate information about the value of companion animal studies to reach a diverse set of researchers. (Shelton)
- Develop funding strategies to enable wider access to resources by the community of researchers engaging in studies of companion animals as sentinels. (Carlin)
- Identify interdisciplinary collaborators who pioneer approaches of use to companion animal studies. (Promislow)

[a] This list is the rapporteurs' summary of points made by the individual speakers identified, and the statements have not been endorsed or verified by the National Academies of Sciences, Engineering, and Medicine. They are not intended to reflect a consensus among workshop participants.

cally the development of effective treatments for some melanomas and lung cancers. He noted that while he served as Acting Commissioner at the U.S. Food and Drug Administration (FDA), one-third of newly licensed therapeutics and devices were cancer related. Nonetheless, social determinants of health,[3] including exposure to environmental carcinogens, can lead to significant disparities in cancer incidence and outcomes. Although biological research tends to focus on understanding the precise mechanism by which a given agent causes a particular cancer, the FDA and the U.S. Environmental Protection Agency (EPA) face a different challenge, which is the practical need to regulate multiple exposures on a daily basis. This is of pressing importance, and the NCI and NIEHS can provide the scientific underpinning needed to help inform these policy decisions, said Sharpless.

The NCI currently funds roughly 80 extramural grants to study environmental impacts on cancer, covering a wide range of topics, said Sharpless. In addition, environmental agents are a focus of larger extramural and intramural studies, including many cohorts within the 20-year-old NCI Cohort Consortium.[4] The NCI supports the monographs program of the World Health Organization's (WHO's) International Agency for Research on Cancer (IARC). The IARC aims to classify environmental carcinogens;[5] recently initiated the Connect for Cancer Prevention study, which follows 200,000 individuals over time with collections of biospecimens[6] and detailed surveys of environmental exposures;[7] and participates in the landmark NIEHS/NCI Agricultural Health study, which monitors the effects of pesticides.[8] Intramural research within the NCI's Division of Cancer Epidemiology and Genetics[9] includes:

- The Diesel Exhaust in Miners Study (DEMS), focused on lung cancer (Attfield et al., 2012);
- The New England Bladder Cancer Study, monitoring the effects of arsenic and other contaminants in drinking water;[10]

---

[3] Social determinants of health are the conditions in the environments where people are born, live, learn, work, play, worship, and age that affect a wide range of health, functioning, and quality-of-life outcomes and risks. See https://health.gov/healthypeople/objectives-and-data/social-determinants-health (accessed March 28, 2022).

[4] See https://epi.grants.cancer.gov/cohort-consortium/cohort_projects.html (accessed February 3, 2022).

[5] See https://monographs.iarc.who.int (accessed February 3, 2022).

[6] Biospecimens are also known as biological samples.

[7] See https://www.cancer.gov/connect-prevention-study (accessed February 3, 2022).

[8] See https://aghealth.nih.gov (accessed February 3, 2022).

[9] See https://dceg.cancer.gov (accessed February 3, 2022).

[10] See https://dceg.cancer.gov/research/cancer-types/bladder/bladder-new-england (accessed February 3, 2022).

- A study examining the impact of coal-fired stoves and heaters on the development of lung cancer in never-smoking women in China;[11] and
- An examination of the effects of pesticides on multiple myeloma (Hofmann et al., 2021).

The NCI also chose environmental carcinogens as one of six topics to focus on for increased cooperation with the United Kingdom. The NCI has extensive interactions with industry, including supporting the development of novel therapeutics and devices, publishing data sets through NCI portals, and funding mechanisms to support the development of new companies, said Sharpless.

A recent NCI meeting, *Breast Cancer and the Environment*,[12] highlighted many promising new tools for studying the cancer–environment connection, including ultrasensitive detection, data aggregation, and exposure monitoring. Referencing the emerging science of mutational signatures, Sharpless envisioned modern, molecularly precise Koch's postulates that could ascribe a particular cancer to a particular agent based on the pattern of somatic DNA mutations in the tumor. James DeGregori (University of Colorado) noted the role played by "selectogens"—chemicals that induce cancer by altering the context around the tumor rather than by mutagenesis. There will likely be many such compounds with a variety of mechanisms, such as the induction of chronic inflammation, and no clear mutation signature, agreed Sharpless. "They are environmental agents that clearly induce cancer . . . but they are not mutagens. They do not damage DNA. They do not break DNA. They do not bind to DNA. They do not have an ascertainable signature," said Sharpless, adding that a scientific approach is needed to study them. Birnbaum suggested standardizing in vitro bioassays and defining a set of key characteristics for evaluating potential carcinogens that do not damage DNA.

New approaches are needed to study the links between environmental agents and cancer, said Sharpless. The time-honored troika of biological plausibility, rodent research, and cohort studies is useful, but many rodent studies have failed to predict outcomes in humans, he said. Companion animals, which share their human's environment, could potentially bridge this gap. The NCI is also very interested in studying cancer immunotherapies in clinical trials for companion dogs with naturally occurring tumors, said Sharpless. Because rodents tend to be poor models for evaluating immuno-oncology agents, these treatments often go straight to human clinical testing, which is very inefficient, and Sharpless expressed hope that dog trials could advance treatments for both

---

[11] See https://dceg.cancer.gov/research/what-we-study/indoor-air-pollution-lung (accessed February 3, 2022).

[12] See https://www.cancer.gov/research/areas/causes/breast-cancer-and-the-environment-controversial-and-emerging-exposures.pdf (accessed December 12, 2021).

animals and humans. The NCI funds a canine immunotherapy trials network[13] and has developed dog-specific antibodies for use in cancer immunotherapy.

The NCI also created a large Integrated Canine Data Commons (ICDC)[14] that maintains data from clinical trials for dogs with cancer. The ICDC is a node in the NCI's Cancer Research Data Commons (CRDC).[15] The ICDC contains a wide variety of genetic, clinical, immunological, and imaging sets in a cloud-based resource containing computational tools, said Amy LeBlanc (NCI), who encouraged researchers to use it for sharing and analyzing data. Researchers can bring their own data and tools to the cloud and combine them with data in the CRDC for integrative analysis, explained LeBlanc.

HISTORY AND CURRENT STATE OF THE SCIENCE OF ENVIRONMENTAL EXPOSURE EFFECTS ON AGING AND CANCER SUSCEPTIBILITY

Speakers offered background on the data linking environmental exposures to cancer and aging in humans and companion animals and considered the relevance of dog genetics to the study of human age-related disease. Speaker presentations focused on the latest science related to understanding the relationship between environmental exposures and cancer and aging in humans and in companion animals, respectively.

Environmental Exposure and Cancer

Cancer Demographics

Cancer is a complex group of over 100 diseases in which cells grow uncontrollably, said Gary Ellison (NIEHS/NCI), who summarized the science linking environmental exposures to cancer in humans. Cancer can occur and spread anywhere in the body, and one tissue can be the source of multiple types of cancer—for example, there are several genetically distinct subtypes of breast cancer, and each could have a different etiology. Approximately, 5 to 10 percent of cancers are due to inherited mutations, with the remaining due to errors that occur during cell division or as a result of DNA damage from the environment.[16] The incidence of most cancers increases with age.

---

[13] See https://dctd.cancer.gov/NewsEvents/20190327_canine_immunotherapy.htm (accessed February 3, 2022).

[14] See https://caninecommons.cancer.gov (accessed December 6, 2021).

[15] See https://datascience.cancer.gov/data-commons (accessed December 15, 2021).

[16] See https://www.cancer.gov/about-cancer/understanding/what-is-cancer and https://www.cancer.gov/about-cancer/causes-prevention/genetics (both accessed March 30, 2022).

Cancer mortality rates have decreased in both males and females in recent years, likely due to advances in detection and treatment (Islami et al., 2021). Nonetheless, melanoma, myeloma, and kidney and liver cancers have been increasing, and environmental factors may play a role. Colorectal cancers have been increasing in those born after 1960, with approximately 10 percent of new diagnoses in individuals under age 50 (Stoffel and Murphy, 2020). The incidence of these cancers is 1.75- to 3-fold higher in those born after 1990 than in 1930, suggesting a possible effect of diet or early life exposure, said Ellison (Figure 1).

Increased incidence for melanoma, colorectal cancer among young adults, liver cancer, and kidney and renal pelvis cancer

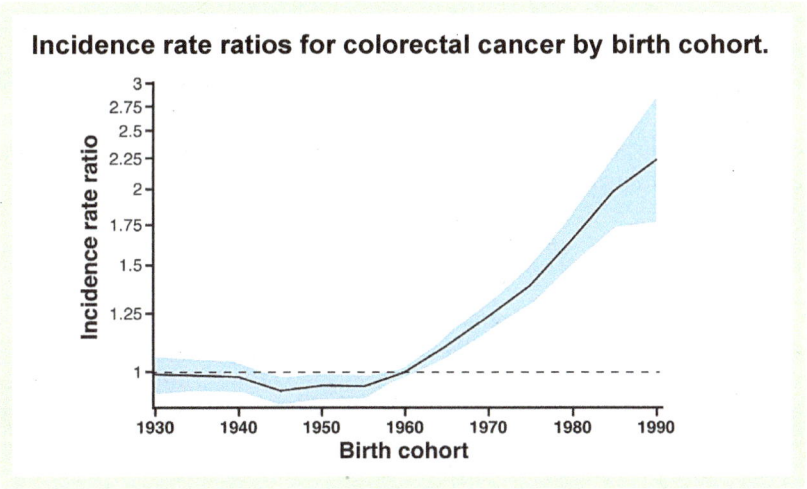

FIGURE 1 Increasing incidence of common cancers and specifically colorectal cancers (CRCs): Environmental etiology?
The incidence of CRC is rising among adults aged 50 years and younger. Ten percent of new diagnoses of CRC are now made among these individuals and disproportionately affect non-Hispanic Blacks, who have nearly twice the rate of non-Hispanic Whites. There appears to be a cohort effect. The incidence rate ratios compare all years to the incidence rate of CRC for those born in 1930.
SOURCES: Ellison presentation, December 1, 2021. Figure is reprinted from Stoffel and Murphy, 2020, and NCI data is from Islami et al., 2021.

Characterizing Environmental Carcinogens

Ellison offered a broad definition of "environment" that comprises all nongenetic factors, including pollutants in air and water; chemicals in building materials, cosmetics, sunscreens, and medicines; pesticides and agricultural chemicals; synthetic additives in food; and lifestyle factors such as nutrition, smoking, and stress. Any of these factors may convey an increased risk of cancer, either by inducing growth-promoting changes in a cell or by altering the cell's milieu (e.g., immunosuppression). Carcinogens are characterized based on the types of genetic, epigenetic, and functional changes they induce in cells (Smith et al., 2016), and exposures to different types of carcinogens lead to different types of cancer. For example, physical exposures such as radiation and ultraviolet light are associated with stomach cancer and melanoma, respectively; pesticides with lung cancer, leukemia/lymphoma, and bladder cancer; diet and exercise with liver, pancreatic, and colorectal cancer; infectious agents with cervical and liver cancer, among others; and reproductive hormones with breast cancer in females and prostate cancer in males. Tobacco smoking is associated with many cancer types.[17]

Two agencies, the NTP[18] and IARC, are tasked with evaluating the carcinogenicity of potentially hazardous substances. As of 2016, the NTP listed 62 known and 186 likely human carcinogens. The IARC has identified 121 human carcinogens, 90 probable carcinogens, and 322 possible carcinogens of over 1,000 agents evaluated.[19] A determination of carcinogenicity is based on evidence that exposure leads to cancer in humans and experimental animals, as well as on mechanistic studies, explained Ellison.

Documenting Environmental Exposures

The complexity of documenting exposure to environmental carcinogens is a particular challenge, said Ellison (Figure 2).[20] The association of asbestos with lung cancer and benzene with leukemia are well-documented examples of occupational exposures to carcinogens; in these cases, exposures were high dose, long in duration, and easy to document. However, many environmental

---

[17] See https://monographs.iarc.who.int/human_cancer_known_causes_and_prevention_organ_site (accessed February 3, 2022).

[18] See https://ntp.niehs.nih.gov/whoweare/index.html?utm_source=direct&utm_medium=prod&utm_campaign=ntpgolinks&utm_term=whoweare (accessed December 16, 2021).

[19] See https://ntp.niehs.nih.gov/ntp/roc/content/introduction_508.pdf (accessed March 30, 2022).

[20] See https://www.epa.gov/risk/guidelines-human-exposure-assessment (accessed December 6, 2021).

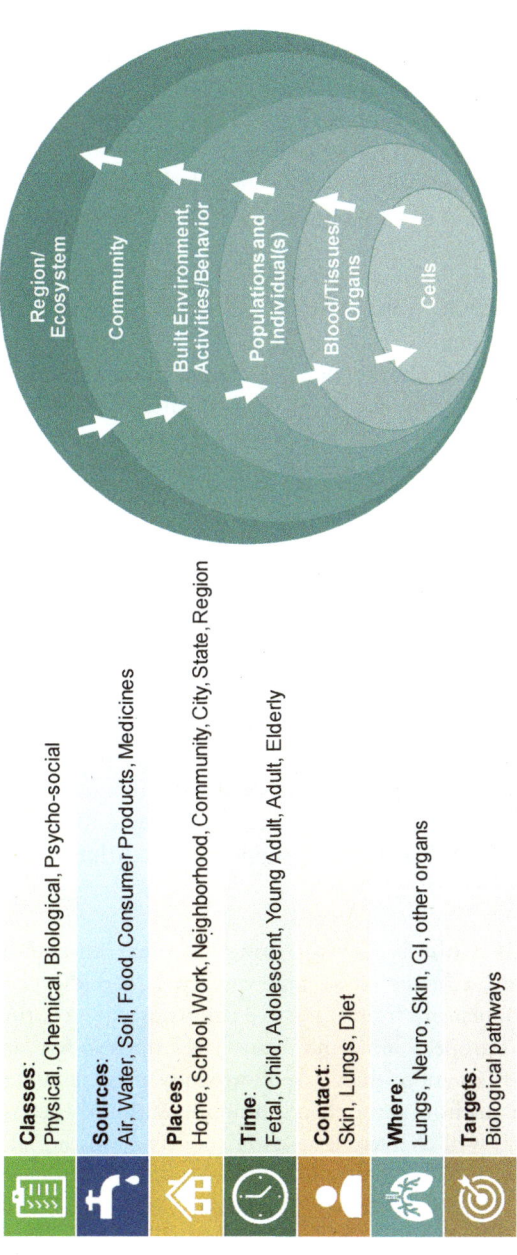

FIGURE 2 The complexities of documenting exposures.
Health and disease are affected by a confluence of the types of exposures, their sources, where and how people are exposed, when in the life span they are exposed, the route of exposure, the pathway through the body, and the effects on biological pathways. Exposures can be measured or modeled at different levels of integration in space and time, from source to dose, and among different biologic, population, and geographic systems.
SOURCES: Ellison presentation, December 1, 2021. Figure courtesy of the National Institute of Environmental Health Sciences.

exposures occur in lower doses in nonoccupational settings and have long latency. Exposures can occur in complex mixtures, with doses and interactions among agents mediating their effects. The timing of exposure is important, as there may be windows of greater susceptibility to both genetic and epigenetic changes across the life span. This is particularly true during gestation and early life, when exposures to environmental toxins may produce lifelong effects. Compounding these difficulties is the fact that retrospective assessment of exposure is not very reliable. Other challenges include variation among individuals in genetic risk factors and windows of susceptibility, tissue-specific effects, and effects that are only observed in subsequent generations, said Ellison.

Cheryl Lyn Walker (Baylor College of Medicine) and Ellison both noted that significant racial disparities exist in cancer incidence and mortality and that these are likely due to racialized social structures, such as segregation in historically redlined neighborhoods, where environmental health disparities are pervasive and affect the quality of food, medical care, and proximity to environmental toxicants. It is also important to consider the impact on cancer of trauma associated with racism, said Ellison. Race is a social construct, he noted, adding that "we need to get away from looking at race and ethnicity, and look at other factors that are associated with racial differences in this country." These factors can lead to epigenetic changes, which are heritable, so the effects of environmental health disparities may be transmitted through the generations, added Walker.

Ellison identified the need to further evaluate probable and possible environmental carcinogens, to evaluate agents linked to cancer in animals when human evidence is insufficient, to consider variations in genetic susceptibility and windows of exposure in prospective study design, to evaluate the cancer risk associated with low levels of agents and with complex mixtures, and to investigate exposures in diverse populations. He described three National Institutes of Health (NIH) efforts to support research into human environmental exposures:

- The NCI Cohort Consortium,[21] an extramural–intramural partnership, consists of 61 high-quality, large cohorts representing diverse populations from at least 15 countries and 4 continents (North America, Europe, China, and Australia). Extensive risk factor data are available from more than 7 million study participants and baseline biospecimens from more than 2 million. Within this consortium, the NCI funds well-characterized aging cohorts consisting of 1.1 million individuals.

---

[21] See https://epi.grants.cancer.gov/cohort-consortium (accessed December 17, 2021).

- A new partnership between the NCI and the NIEHS[22] recently funded five new cohorts to study the effects of emerging and important environmental exposures on cancer risk. These cohorts are racially/ethnically diverse, will generate data on a wide variety of exposures, and will enable the study of associations between exposures and biomarkers such as immune modulation, oxidation, hormone-mediated effects, and cell immortalization.
- The Human Health Exposure Analysis Resource: 2019–2024 (HHEAR),[23] led by the NIEHS, supports investigators who want to add or expand exposure analysis to their human health research. The HHEAR offers three areas of support: a network of laboratories to analyze biological and environmental specimens using both targeted and untargeted approaches, a coordinating center, and a data center. The HHEAR currently has more than 40 studies in its data repository, said Yuxia Cui (NIEHS), and all the exposure data and associated phenotypic information will eventually be made publicly accessible to researchers.

Environmental Exposure, Cancer, and Aging in Companion Animals: The Companion Dog Model

Dogs as Sentinels for Human Disease Agents: Shared DNA, Environment, and Disease

The utility of animal sentinels for environmental exposures is illustrated by the classic example of "the canary in the coal mine", where canaries were brought into mines early in the 20th century to sense carbon monoxide (CO) exposure, said Peter Rabinowitz (University of Washington).[24] Canaries displayed the three qualities of an animal sentinel needed for assessing the health risks posed to humans by environmental hazards: greater susceptibility to the hazard; greater exposure (in the case of canaries, more rapid metabolism of CO); and shorter latency.

The usefulness of dogs and cats as sentinels of harmful environmental exposures has been demonstrated through several public health calamities over

---

[22] See https://grants.nih.gov/grants/guide/rfa-files/RFA-CA-20-049.html (accessed December 17, 2021).

[23] See https://grantome.com/grant/NIH/U24-ES026539-02 (accessed December 17, 2021).

[24] See https://www.smithsonianmag.com/smart-news/story-real-canary-coal-mine-180961570 (accessed December 15, 2021).

the last century, said Kurunthachalam Kannan (New York University School of Medicine). A classic example arose from industrial mercury dumping in the fishing village of Minamata, Japan, which contaminated the local fish supply; local cats, who ate leftover fish, displayed "dancing cat fever" years before the neurological symptoms of mercury poisoning became evident in people (James et al., 2020). The deliberate addition of melamine to pet food in 2007 caused kidney failure in cats and dogs, and its addition to infant formula in China the following year caused kidney failure in children (Skinner et al., 2010); thousands of children were affected and several died. In the 1940s and 1950s, the high rate of bladder cancer in dogs exposed to aromatic amines linked these compounds to the high rate of bladder cancers in textile industry workers (Dietrich and Golka, 2012; Hayes et al., 1981).

The etiology of cancer involves both genetic and environmental factors, said Audrey Ruple (Virginia Tech), so the best sentinels for carcinogens would be similar to humans in terms of both genetics and exposure. Dogs share about 650 million base pairs of ancestral DNA sequence with humans, and dog orthologous genes are more similar to humans than are the same genes from mice (Kirkness et al., 2003; Paoloni and Khanna, 2008). The convergent evolution of dog and human genes is likely due to similar selective pressure from shared environmental exposures over many millennia, said Ruple. The geological record documents human exposures to dust and silica going back millions of years, joined by wood smoke roughly a million years ago, both of which left an imprint in the human genetic code, said Caleb "Tuck" Finch (University of Southern California) (Trumble and Finch, 2019). Domestication of dogs roughly 20,000 years ago may have brought them into the smoky environment of small human dwellings of the time, said Finch. Shared exposures are still the norm in the 40 percent of U.S. homes that include pet dogs, said Ruple. Not only do pets share a common living space with humans, but their behavioral patterns resemble those of toddlers, including mouthing objects and contact with the ground, noted Kannan. Dogs are a fantastic model species from a genetics standpoint, added Adam Boyko (Embark Veterinary, Inc./Cornell University). There are hundreds of dog breeds and they are much more differentiated than human populations, with larger ranges of phenotypic diversity, including a 50-fold difference in body size that correlates with a 2-fold difference in aging rate. This provides a useful model for genetic association studies and for mapping the mutations that underlie specific phenotypes, said Boyko.

Sharpless, Ruple, Rabinowitz, and Kannan listed multiple advantages to using companion dogs as a model for environmentally induced cancers, many of which are summarized in Figure 3 from Sharpless's presentation. Ruple emphasized the difficulty in identifying environmental carcinogens in humans due to a long latency period, whereby decades may elapse between exposure and disease. In contrast, dogs are short lived and their cancers have

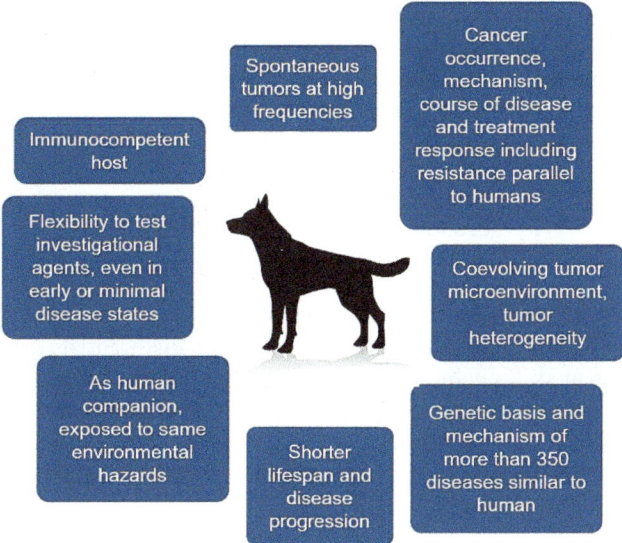

FIGURE 3 Dogs can serve as a useful model in comparative oncology.
Dogs and humans share the same environment and possess many similarities in genetics, physiology, tumorigenesis, and cancer progression, with the shorter life span of dogs corresponding to more rapid disease progression.
SOURCES: Sharpless presentation, December 1, 2021. Adapted from Bujak et al., 2018.

a much shorter latency. Matthew Breen (North Carolina State University) noted that the roughly 7-fold difference in longevity between humans and dogs translates into potential differences in latency of exposure, said Breen, allowing scientists to "harness the impacts of environmental influences on our pet dogs, living in our environment but over a much shorter period of time." Rabinowitz and Kannan both highlighted the example of mesothelioma, which is largely caused by asbestos exposure and has a latency of 30–35 years in humans, with a mean age of about 8 years in dogs (Glickman et al., 1983). Ruple noted that the incidence of cancer is 10 times higher in dogs than humans. She added that studying the tumors that develop spontaneously in companion dogs (Schiffman and Breen, 2015) can avoid the ethical concerns associated with inducing tumors in laboratory animals. Rabinowitz added that some breeds have increased susceptibility to certain cancers, such as osteosarcomas in golden retrievers and histiocytic sarcomas in Bernese mountain dogs, and might therefore be more sensitive to certain carcinogens. Many studies have demonstrated the utility of dogs as sentinels for the reproductive effects of environmental contaminants (Sumner et al., 2020). Richard Lea (University of Nottingham) commented that dog testes and ovaries, which are disposed of in large quantities following spay/neutering procedures, are a valuable resource that can be used both to screen for environmental contaminants and to study the connection of these agents to specific pathologies. Breen emphasized that the different types of cancer shared between humans and dogs show "a huge amount of similarity" in both anatomic pathology and genetics (Figure 4).

Why Pet Dogs with Spontaneous Tumors are Good Models for Human Disease

"What is very striking is the sheer number of dogs that get cancer each year in the United States and the overall incidence that translates into," added Breen. Out of the roughly 60 million pet dogs that visit a veterinarian for health care each year in the United States, 6 million are diagnosed with cancer. "This represents an enormous opportunity to provide health data and biological specimens from dogs over time and geographical space over the United States," said Breen.

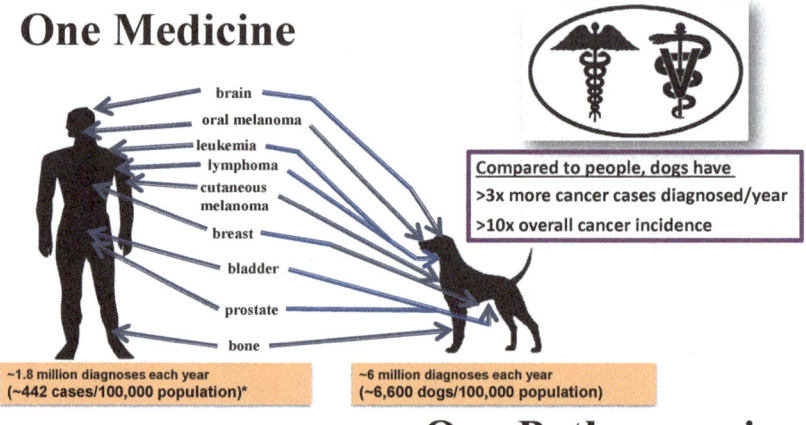

FIGURE 4 The remarkable similarity in cancers that develop in humans and dogs. The cancers shown are found in both human and canine populations, and several ongoing studies are identifying similarities at the genomic level. The incidence of cancer in both species is shown, highlighting the more than 3-fold higher number of cancers diagnosed in pet dogs each year, as well as a 10-fold overall cancer incidence. Comparative oncology is a growing transdisciplinary field that harnesses these data, adding evidence to understand a shared pathogenesis.
SOURCES: Breen presentation, December 1, 2021. Figure adapted from Schiffman and Breen, 2015. * Data from the NCI, www.cancer.gov (accessed December 1, 2021).

Using advances in data collection, researchers have recently begun to develop and analyze large veterinary data sets (Paynter et al., 2021). These include prospective longitudinal studies in dogs, such as the Golden Retriever Lifetime Study (Box 3) and the Dog Aging Project (Box 4); see section on Ongoing Canine Population Studies.

Ruple noted several limitations to the use of animals as sentinels of environmental exposures. There is no nationwide census of dogs in the United States and no dog cancer registry, which makes it difficult to assess the actual incidence of cancers in dogs. The absence of information regarding the numbers and types of pets in U.S. households was referred to throughout the meeting as "the denominator problem." Because dogs do not receive death certificates, the burden of cancers remains unknown and is not captured by veterinary records that list the cause of death as euthanasia. The failure to record the reason why these dogs are being euthanized is a flaw in the current system, said Ruple.

Ellison noted that companion dog research lends itself well to citizen science approaches and that the NIEHS is a leader in this approach, with most funding opportunities now requiring community engagement. Citizen scientists collect biospecimens like toenails, hair, and teeth from their pets to capture data on exposures. Ruple cautioned that there are considerable health disparities in veterinary medicine, with the enrolled dogs primarily from mostly affluent White people, in research studies aimed at investigating environmental impacts on cancer that she has been involved with. These are the same individuals already represented in the existing data sources. These disparities in participation need to be understood as "a threat to the validity of the work we are doing" and actively addressed by engaging communities that are not well represented. Otherwise, this research will only exacerbate the inequities that already exist in human medicine, warned Ruple. Ellison raised the possibility of leveraging data from existing human cohort studies supported by the NCI to include dogs to increase representation for both human and dog research.

Animals as Sentinels of Environmental Health Hazards: A Brief History

Ruple briefly reviewed the findings from a 1991 NASEM consensus report entitled "Animals as Sentinels of Environmental Health Hazards," which focused on efforts to monitor cancer-related health outcomes in dogs, primarily through retrospective, observational case-control studies assessing exposures in healthy and sick animals (NRC, 1991). One such study found an 8-fold increase in dog mesothelioma if the owner was exposed to asbestos (Glickman et al., 1983), while others found positive associations between a range of cancers in dogs and the use of herbicides or pesticides. Ruple noted that studies on the carcinogenic effects of pesticide use in humans, which have begun to appear in the last 5 years, took 20 years to complete, whereas the dog studies covered in the 1991 report were done in considerably less time.

Studies published since that 1991 report have demonstrated further associations between cancer in companion dogs and environmental pollutants, as well as between dog cancer and smoking by the owner. Search-and-rescue dogs deployed to the World Trade Center following the 9/11 attacks did not have statistically significant increased cancer rates compared to controls 15 years later, which could offer reassurance to humans who worked on the site (Otto et al., 2020). Studies cited by Ruple and Rabinowitz in their presentations are summarized in Table 1. Additional canine and feline cancer sentinel studies are reviewed in Ruple et al. (2019). Summarizing the history of canine cancer sentinel studies, Rabinowitz noted that studies of dogs as sentinels have untapped potential for detecting environmental carcinogens and improving human cancer prevention.

Lessons Learned, Looking Forward

Following the Flint, Michigan, water crisis that poisoned thousands of children with lead, it was found that local dogs also had higher levels of lead in their blood than unexposed controls (Langlois et al., 2017). Earlier studies in the 1970s had shown that dogs could be used to detect areas with problematic levels of lead paint, and thus address the environmental health needs of underserved populations, but this work was abandoned in the late 1970s, said Ruple. She added that if there had been an ongoing effort to identify lead levels in dogs this could have made a difference: "Imagine if we had not stopped … instead of waiting until more than 14,000 children were impacted in Flint, Michigan."

The 1991 NASEM consensus report showcased a wealth of studies demonstrating the utility of using animals as sentinels for environmental health risks, concluding that "these findings indicate that well-designed epidemiological studies of spontaneous tumors in pet animals may provide insight into the role of environmental factors in human cancers and serve as a valuable sentinel model to identify environmental health hazards for humans" (NRC, 1991). But it also acknowledged that this approach had not gotten much traction. "One reason," it offered, "might be the institutional inertia that accompanies integration of new scientific methods into the risk-assessment process and use of the results for risk management. Many government agencies do not recognize the importance of animal sentinels or agree on how to compare the findings obtained with them and the findings obtained with more traditional methods." The report noted that research in animal sentinels was not a high priority for funding—however, it recognized that increasing interest in

TABLE 1 A Selection of Canine Cancer Sentinel Studies Showing Positive Association or No Association with Environmental Exposure

Studies Showing Positive Association with Environmental Exposure

Year	Findings	Citation
1983	Asbestos bodies detected in 3 of 5 dogs with mesothelioma	Harbison and Godleski, 1983
1983	Case-control study (n = 16): Owner exposure to asbestos associated with mesothelioma in dogs	Glickman et al., 1983
1989	Exposure to topical insecticides associated with bladder cancer in dogs	Glickman et al., 1989
1991	Cases of canine malignant lymphoma (n = 491) associated with exposure to 2-4D lawn chemicals	Hayes et al., 1991
1992	Passive smoking associated with lung cancer in short-nosed dogs	Reif et al., 1992
1998	Passive smoking associated with nasal cancer in long-nosed dogs	Reif et al., 1998
2001	Canine lymphoma associated with living in an industrial neighborhood	Gavazza et al., 2001
2004	Dog transitional cell carcinoma associated with both insecticide and herbicide use, with an additive effect	Glickman et al., 2004
2009	Canine lymphoma cases (n = 608) associated with exposure to waste incinerators, polluted sites, and radioactive waste	Pastor et al., 2009
2012	Canine lymphoma cases (n = 263) associated with commercial lawn pesticide use	Takashima-Uebelhoer et al., 2012
2017	Geographical variation found in incidence of golden retriever lymphoma (n = 454)	Ruple et al., 2017
2018	Gut microbiome of dogs with lymphoma (n = 12) differs from that of healthy dogs	Gavazza et al., 2018
2020	Boxers with lymphoma (n = 63) more likely to live near a nuclear power plant, chemical suppliers, or crematorium	Craun et al., 2020

continued

TABLE 1 Continued

Studies Showing Positive Association with Environmental Exposure

Year	Findings	Citation
2021	Urothelial cell cancer in dogs ($n = 63$) associated with counties with higher water trihalomethanes and air ozone levels. Also found lymphoma in boxers associated with counties with higher ozone and airborne 1,3-butadiene and formaldehyde	

Studies Showing No Association with Environmental Exposure

Year	Findings	Citation
2008	No significant association between dogs with bladder cancer ($n = 100$) and trihalomethane in water	Backer et al., 2008
2020	15-year follow-up of World Trade Center rescue dogs ($n = 95$) showed more particles in lungs but no increased cancer compared to controls	Otto et al., 2020

SOURCES: Adapted from Ruple and Rabinowitz presentations, December 1, 2021.

humane alternatives to lab animal experimentation might draw attention to the use of companion animals, as indeed has transpired.

Workshop participants considered how data on animal exposures could be used to advance public health. Ruple said that work needs to be done on both the human and animal sides to bridge the gap, and noted that veterinarians are often the first ones to sound the alarms on vector-borne disease. To bring the One Health[25] voice to the public health arena, she emphasized that veterinarians need to be included on public health boards, which is happening in some places. She added that updates are needed in the curriculum of training programs. Marcia Haigis (Harvard University) and Elaine Ostrander (National Human Genome Research Institute) emphasized the need to make the large canine '-omic data that is now being accumulated freely accessible to investigators so they can use informatics approaches to generate hypotheses for cancer mechanisms, as is done for humans. Ostrander applauded the efforts of the NCI's canine commons, but added that the problem of access will only be solved as a community. Rabinowitz envisioned a future when veterinarians could detect an environmental hazard in a companion animal and convey this information to the medical professionals caring for members of the household. "There needs to be a way to clinically get that information to people concerned with human health . . . I really hope we move in that direction," he added.

How Diet Modulates the Tumor Microenvironment (TME)

Metabolites and the TME

Obesity is a risk factor for several types of human cancer (Lauby-Secretan et al., 2016), and this is becoming increasingly significant as our metabolic health demographics change, with nearly 50 percent of the U.S. population expected to be obese by 2030 (Ward et al., 2019), said Marcia Haigis (Harvard University). She described her efforts to understand the connection between diet, a type of environmental exposure, and cancer. Extending the central dogma to metabolites (DNA→RNA→protein→metabolite), Haigis noted that 25 percent of the proteome has a direct metabolite readout, and that this readout is directly affected by both the external environment and diet. Metabolites feed back to the earlier steps of the central dogma, controlling protein function through allosteric effects and posttranslational modification, and influencing DNA packaging, with consequences for epigenetics and gene expression.

---

[25] See https://www.cdc.gov/onehealth/index.html (accessed February 11, 2022).

Cancer cells in particular have tremendous energy demands and burn fuel by metabolizing glucose, fatty acids, and amino acids, generating profiles of metabolic byproducts distinct from those of healthy cells. For example, tumor cells from estrogen-receptor-positive (ER+) breast cancers accumulate large amounts of ammonia, which they secrete into the cellular environment as a metabolic waste product but then recycle and use for building amino acids, unlike healthy cells (Spinelli et al., 2017). Tumor metabolites can alter protein activity through mechanisms similar to those used by the metabolites of healthy cells. But tumor cells do not exist in isolation, said Haigis; their growth and survival are mediated by interactions with a host of cells and molecules within the tumor microenvironment (TME). Although killing by cytotoxic T lymphocytes (CTLs) is important for controlling tumors, tumors have evolved several mechanisms for evading CTLs, including by constraining the ability of CTLs to function in the TME, which can be accomplished by sequestering fuel away from the CTLs or by secreting metabolites—such as ammonia and lactate—that inhibit CTLs.

Diet and the TME

How does diet affect cancer risk? Haigis described highlights of a recently published study that investigated how a high-fat diet alters the function of CTLs and tumor cells, using implanted tumors to probe tumor-driven modification of the TME in mice (Ringel et al., 2020). Syngeneic[26] colorectal tumor cells were implanted into mice fed either a control or high-fat diet, and tumor growth and phenotypes were monitored over time. When mice were fed a high-fat diet, CTLs were suppressed in the TME and the incidence and growth of tumors increased. $CD8^+$[27] CTLs in particular were fewer, less proliferative, and less cytotoxic in the TME of mice fed a high-fat (vs. control) diet. In addition, $CD8^+$ CTLs from the TME of mice on the high-fat diet changed their choice of energy source, up-regulating fatty-acid metabolism and down-regulating glucose metabolism.

Drilling down to the molecular level, Haigis and colleagues found that expression of the prolyl-hydroxylase 3 (PHD3) protein was down-regulated in tumor cells exposed to a high-fat diet. PHD3 belongs to a family of enzymes that mediate the response to hypoxia stress. PHD3 represses fat metabolism (German et al., 2016), so down-regulation of PHD3 would be expected to increase fat burning and oxidation. The researchers found that, although the high-fat diet increased fatty acid metabolites (lipids) in circulation, as might

---

[26] Syngeneic is defined as genetically similar or identical.
[27] CD8-positive T cells are a critical subpopulation of major histocompatibility complex (MHC) class I-restricted T cell and are mediators of adaptive immunity.

be expected, it had the opposite effect in the TME, where it caused lipid accumulation to decrease, potentially because the tumor cells, having turned off their PHD3 gene, were taking up all the lipids. Suspecting this would have the effect of inhibiting CTLs by depriving them of energy from lipid metabolism, the researchers induced the tumor cells to overexpress PHD3. In these mice, lipid levels in the TME remained high even on a high-fat diet, CTL numbers increased, and tumor growth was reduced.

Taking this question to human studies, Haigis and colleagues found that PHD3 expression was similarly decreased in tumors from colorectal cancer patients with a high body mass index (BMI) compared to low-BMI patients, and these PHD3-low tumors were immunologically "cold," with low levels of tumor-infiltrating lymphocytes, compared to PHD3-high tumors. Haigis concluded that obesity metabolically reprograms tumors to metabolize fat while changing the TME and impairing antitumor immunity—and this immunity may be restored by manipulating the expression of a single gene. Haigis emphasized the importance of studying the effect of environmental factors and aging on tumor growth on the cellular level. She also noted that most laboratory models of cancer use young animals, despite the importance of aging in cancer development.

Epigenetic Aging as a Target and Biomarker for Environmental Exposures

Environmental Exposures Alter the Epigenome

The epigenome, which consists of heritable chemical modifications of DNA, without changes to the sequence of bases, or chromatin, is a target for environmental exposures, said Cheryl Lyn Walker (Baylor College of Medicine). The primary type of epigenetic modification is methylation of CpG dinucleotides in the DNA sequence, and this can be inherited mitotically or meiotically. In contrast to DNA, which is protected from environmental influences by multiple mechanisms, the epigenome is inherently plastic, and it is well equipped to respond to the environment (Perera et al., 2020). One example would be epigenetic modifications in a fetus in response to starvation, which are adaptive for development in a nutrient-poor environment following birth, though they can lead to metabolic disease in an environment that is nutrient rich (Fleming et al., 2018). Epigenetic modifications may persist long after an exposure has occurred and can occur late in life. For example, breast tissue is developmentally plastic until the first full-term pregnancy (Walker and Ho, 2012).

Epigenetic modifications to DNA continue to change in species-specific patterns over time as the organism ages, and this predictable course of DNA methylation has enabled the characterization of species-specific biological clocks (Field et al., 2018). The biological clock runs at different speeds in different individuals, revealing differences in "biological age" between humans of the same chronological age. These differences in biological clocks are excellent predictors of health, said Walker. Environmental perturbations can cause a change in the epigenetic clock, in which a "faster" epigenetic clock predicts decreased survival (Kochmanski et al., 2017), as was demonstrated over a 15-year time period for older males participating in the Veterans Health Administration Normative Aging Study (Belsky et al., 2020). Low socioeconomic status and high trauma were associated with faster clocks (Schmitz et al., 2021; Verhoeven et al., 2018). Epigenetic clocks can be compared across tissues and species. For example, although the maximum life span is 122 years for humans and 3.8 years for rats, recalibrating both species to the same scale shows 95 percent correlation in epigenetic aging between the two (Horvath et al., 2020). These correlations support study of the epigenome as a biosensor in companion animals, said Walker.

Epigenetic Modifications can Enhance Cancer Risk

Walker described her group's work in the TaRGET-II consortium,[28] which is probing the ways in which multiple environmental perturbations alter the epigenome across multiple tissue types. When male mice were exposed in utero to the obesogen tributyltin (TBT), roughly half developed liver tumors by 10 months (Katz et al., 2020). In mice that developed tumors, researchers were surprised to discover that the transcriptional profile of the normal-appearing, uninvolved part of the liver looked nearly identical to that of the tumors. In contrast, the transcriptional profile of the normal liver from TBT-exposed animals that did not develop tumors resembled that of untreated control mice. Walker and colleagues hypothesized that TBT treatment had generated a distinct epigenetic "endotype" in the livers of half the exposed mice, which conferred a high risk for tumor development. Epigenetic analysis of these mice detected many differences in CpG[29] methylation in the uninvolved livers of TBT-treated mice that had developed tumors, compared to livers of untreated control mice, while TBT-treated mice without tumors resembled the controls. These methylation differences were observed in genes that normally underwent epigenetic modifications early in development, and they could be observed in the high-risk mice months before they developed any tumors. Walker surmised that TBT targeted this "plasticity of epigenetic aging," and as a result, normal epigenetic aging failed to occur in the animals

---

[28] See https://target.wustl.edu/index.html (accessed December 15, 2021).
[29] A CpG is a C (cytosine) base followed immediately by a G (guanine) base.

that had been transformed to the high-risk type. Importantly, humans with liver cancer or non-alcoholic fatty liver disease displayed the same high-risk epigenetic signatures associated with TBT-induced reprogramming in mice, indicating that the same genes were altered in the human disease state. The epigenome is an incredible biosensor of the environment, said Walker, especially for analyzing the effects of complex exposures, and with the availability of two-species clocks, the faster epigenetic clock of companion animals presents a valuable tool for inferring the future health of the human epigenome and perhaps for predicting adverse outcomes due to environmental exposure; specific companion animal examples are provided in the next section.

Asked to consider differences in methylation patterns across different cell types, Walker replied that the TaRGET II consortium has not detected any correlation between epigenetic signatures in blood cells and epigenetic changes in target tissues. For example, blood cells failed to distinguish between individuals with different liver epigenome types following TBT treatment. Haigis cited her metabolic analyses, which detected a different abundance of fat metabolites in the plasma versus the tumor microenvironment. This is particularly relevant to epigenetics, she said, because the acetyl-CoA from fat oxidation is used in the acetylation of histones in chromatin, another epigenetic change that can alter gene expression, and there are many other metabolites that regulate enzymes. Different places in the body have different nutrient pools and react differently to changes in diet, said Haigis, and therefore, epigenetic differences between circulating lymphocytes and the cells in tissues should not come as a surprise.

Domestic Dogs as a System for Understanding Aging and Life Span

Development of Dual-Species Epigenetic Clocks

Historically, the epigenetic clocks in humans have not been strong predictors of age in other mammals, and this has presented a challenge to advancing dog studies of aging and cancer. To determine whether the epigenetic age of a dog could predict its life span (as it does in humans), Ostrander and colleagues characterized the methylomes (consisting of 50,000 CpG sites spanning the genome) of blood cells from 104 Labrador retrievers ranging in age from 2 months to 16 years (Wang et al., 2020). Young dogs showed high methylome similarities to young humans, and aged dogs to older humans, with a weaker correlation in midlife. Mapping the methylation patterns against age, the researchers detected a nonlinear relationship between dog and human age; following this formula: human age = 16(ln(dog age)) + 31 (see Figure 5). When life stages were plotted along this curve, the researchers found a strong agreement between the approximate times at which dogs and humans experience physiological mile-

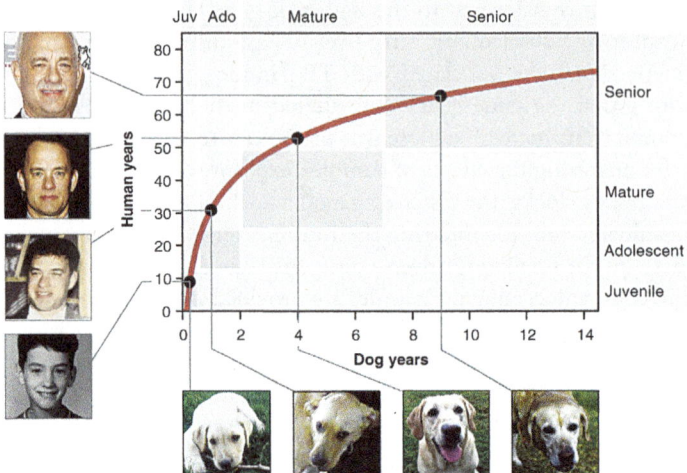

FIGURE 5 The nonlinear relationship between dog age and human age.
Logarithmic function for epigenetic translation from dog age (*x*-axis) to human age (*y*-axis). Outlined boxes indicate the approximate age ranges of major life stages as documented qualitatively based on common aging physiology. Juvenile refers to the period after infancy and before puberty, 2–6 months in dogs, 1–12 years in humans; adolescent refers to the period from puberty to completion of growth, 6 months to 2 years in dogs, approximately 12–25 years in humans; mature refers to the period from 2–7 years in dogs and 25–50 years in humans; senior refers to the subsequent period until life expectancy, 12 years in dogs, which correlates with 70 years in humans. Black dots on the curve connect to images of the same yellow Labrador taken at four different ages (courtesy of Sabrina and Michael Mojica, with permission) and to images of a representative human at the equivalent life stages in human years (photos of Tom Hanks drawn from a public machine-learning image repository, Chen et al., 2015).
SOURCES: Ostrander presentation, December 1, 2021. From Wang et al., 2020.

stones in the transitions from infancy to juvenile, adolescent, mature, and senior life stages. The researchers also observed conservation of epigenetic progression between dogs and mice, suggesting that changes in methylation with age extend from humans to other mammalian genomes, said Ostrander.

Among the 50,000 methylation sites, Ostrander and colleagues identified 394 genes whose methylation patterns showed conserved time-dependent behavior across dogs, mice, and humans. When these genes were mapped onto the PCNet molecular interaction database,[30] which captures physical and functional relationships among genes and their products, they clustered into

---

[30] PCNet is a database of molecular interactions capturing physical and functional relationships among genes and their products.

five highly interconnected network modules involved in anatomical development, leukocyte differentiation and metabolism, neuroepithelial cell differentiation, and synapse assembly and regulation. Some of these genes are associated with cancer as well as aging. Unlike age predictions based on the 50,000-site whole methylome clock, which only worked within the same species used to create the clock, patterns of CpG methylations in these 394 conserved developmental gene modules could make accurate age predictions across multiple mammalian species. "This is allowing us to hone in on the most relevant genes … for building a really good epigenetic clock," explained Ostrander.

Aging, Somatic Evolution, and Cancer—The Inexorable Link

Adaptive Oncogenesis: Why Does Cancer Increase with Age?

Ninety percent of human cancers develop after the age of 50, noted DeGregori, and this has traditionally been ascribed to the age-dependent accumulation of mutations. But he emphasized that the story is more complicated because different cancers are driven by different numbers of mutations, initiate in different stem cell pools, and have different age-dependent incidences. However, Rozhok and DeGregori (2019) found that when curves of cancer incidence are normalized based on their maximums, diverse cancers all demonstrate the same age-dependent increase in incidence over time, regardless of stem cell source or the number of driver mutations required for their development (see Figure 6). DeGregori noted that the likelihood of developing cancer and dying after age 50 is not linear; it gains speed over time, reflecting a dramatic drop-off in natural selection against cancer beyond the age where most individuals would have historically been dead by other causes. This derives, he said, from a variety of cellular and genetic mechanisms that provide resiliency to tissues throughout the reproductive years but decline thereafter, increasing vulnerability to a host of diseases, cancer among them.

The significance of these resiliency mechanisms in suppressing cancer is seen when humans age and the waning of tissue maintenance enables cancers to develop, through a process DeGregori termed "adaptive oncogenesis." Adaptive oncogenesis predicts that an oncogenic mutation in one cell will reduce its fitness relative to the rest of the pool if that pool is healthy but will not incur a selective disadvantage if the pool is damaged (DeGregori, 2011). DeGregori and colleagues tested this theory by transplanting a small number of Ras-transformed stem cells into both young and old mice (Henry et al., 2015). The transformed cells expanded only in the old mice, resulting in leukemias; however, blocking inflammation prevented both expansion and leukemias. DeGregori observed that youth is intrinsically tumor suppressive, with healthy tissues exerting stabilizing selection against clonal expansion of transformed cells. As individuals age and their tissues become damaged, onco-

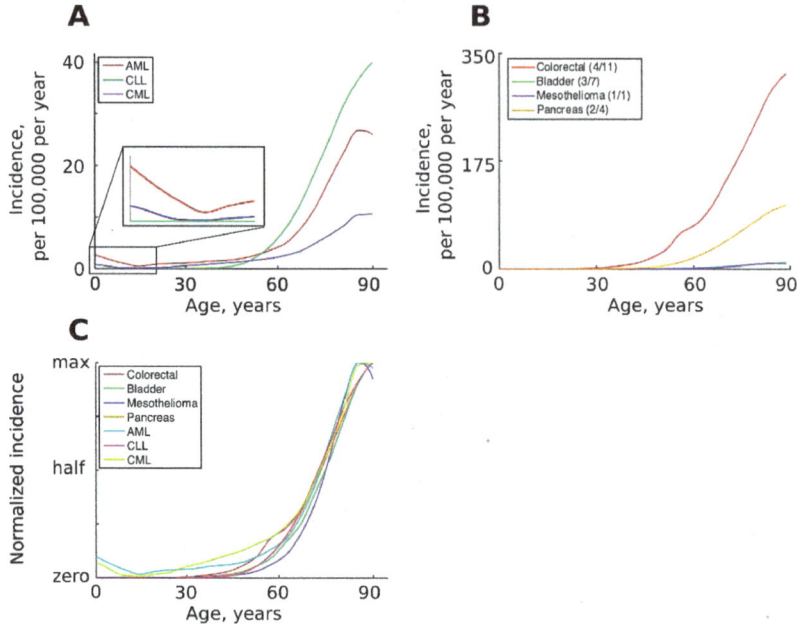

FIGURE 6 Cancers requiring different numbers of driver mutations and originating from stem cell pools that are organized in vastly different ways demonstrate very similar age-dependent incidence. (A) The age distribution of the three most common types of leukemia in humans: AML—acute myeloid leukemia, CLL—chronic lymphocytic leukemia, CML—chronic myeloid leukemia. (B) The incidence of four other cancers; the first number in the brackets indicates the average number of predicted driver mutations in known cancer genes and the second is the average number of predicted driver mutations in all protein coding genes according to Martincorena et al. (2017). The four cancers were chosen for their variability in the predicted numbers of driver mutations. (C) The incidence of cancers from panels A and B normalized by dividing each data point by the corresponding cancer's maximum incidence (removing scale and preserving shape). (Data from National Cancer Institute, www.seer.cancer.gov.)
SOURCES: DeGregori presentation, December 1, 2021. From Rozhok and DeGregori, 2019.

genic alterations that were maladaptive in a young environment are selected for in the older environment, and the incidence of cancer over time is shaped by the changing, age-dependent balance of genetic drift, stabilizing selection, and positive selection (for malignant growth).

How Improving the Function of Aging Tissue May Reduce Cancer Risk

In a computational study, Evans and DeGregori (2021) found that human tissues become "riddled" with cells carrying cancer-associated mutations as they age, with over 100 billion cells in the body harboring at least one known oncogenic mutation by age 60. Roughly 40 percent of adults develop cancer, which arises from a single cell; this raises the question of why the other 100 billion cells with "oncogenic" mutations do not result in cancers. One clue may be found in a study of lung cancer, which demonstrated that, although smoking dramatically increases the likelihood of getting lung cancer at any age, lung cancer incidence shows a similar age-dependent increase in both current smokers and never-smokers (Samet et al., 2009).

To investigate how smoking status influences oncogenesis in the lung, DeGregori and colleagues examined normal (noncancerous) lungs in current, former, and never-smokers. Using duplex DNA sequencing, which enables detection of mutations with high sensitivity, the group found that even a small brushing of lung tissue from a 70-year-old contained hundreds of mutations, dozens of which were cancer associated, regardless of the person's smoking history. However, smoking status influenced proliferation of the cells carrying oncogenic alleles. Cells with cancer-associated mutations demonstrated relatively little expansion in nonsmokers, compared to former and (even more) current smokers. Focusing on the p53 tumor suppressor gene, the highest frequency of cancer-associated alleles was seen in lungs of current smokers, more than former smokers. While these results clearly indicate that smoking promotes selection for oncogenic variants in the lung, they also suggest that quitting smoking may reverse this selection and thus explain why every year that passes after a person quits smoking reduces their odds of getting lung cancer (Tindle et al., 2018), said DeGregori. He added that they also support the goal of developing interventions to accelerate the restoration of more normal lung landscapes after quitting. In a broader sense, these observations indicate that efforts to prevent and treat cancer could converge with similar efforts to prevent other aging-associated diseases, he said.

Discussion: Animal Size, Reproductive Status, and Cancer

The Paradox of Large Animals and Cancer

Daniel Promislow (University of Washington) raised the question of Peto's Paradox,[31] which refers to the surprising fact that large animals like whales and elephants do not die of cancer when they are a few weeks old, given the number of cells that could potentially develop cancer compared to a mouse; rather, the larger species have longer lives. That large dog breeds

---

[31] Sir Richard Peto noted that although mice have approximately 1,000 times fewer cells and > 30 times shorter life spans than humans, their risk of carcinogenesis is not markedly different—this is coined as Peto's Paradox. For more information, see Peto (2015, 2016) and Callier (2019).

develop cancer more frequently than small breeds is the "paradox of the paradox," said Promislow. DeGregori noted that, similar to dogs, big humans have higher cancer rates than smaller humans. The association of larger body size with increased cancer in dogs and humans is likely due to modulation of the insulin-like growth factor (IGF) pathway, which, when hypomorphic,[32] consistently leads to a longer life span and less cancer in model organisms, said DeGregori. Selection for life span versus body size within a species therefore reflects differential investment, with natural selection pitting early reproduction and robustness against long-term maintenance and extended reproductive capacity. Ostrander cited data from her laboratory indicating that IGF alleles associated with small, longer-lived breeds were present heterozygously in wolves 50,000 years ago, roughly 30,000 years before domestication, but did not achieve homozygosity until about 10,000 years ago, when domestication and agrarian society were well established.

Reproductively Altered versus Intact Animals

Participants noted that reproductively altered animals might display different sensitivities to carcinogens than humans, owing to their different hormonal context. Ostrander noted that prostate cancer is a nearly ubiquitous disease of aging in men, except in those who lose their gonads early in life and never develop this type of cancer. In dogs, it is the opposite—prostate cancer is rare but is much more likely in neutered than intact dogs.

The question of longevity of altered versus non-altered companion animals is complicated. Ruple remarked that in the United States, intact animals typically live shorter lives, but they tend to be killed by cars and other traumas. Ostrander added that altered animals in the United States are more likely to be obese, which confounds the effect of the hormonal changes on cancer and mortality. However, spaying and neutering is not consistent worldwide, and research is being done in Hong Kong and other locations, where most dogs and cats are not altered but nonetheless live long lives, said Ruple.

Validation of Canine Models as Sentinels

A fundamental question, said Rod Page (Colorado State University), LeBlanc, and others, is how to validate a canine model as a sentinel. Dog cancer is not always directly comparable to human cancer, and many times it is not possible to obtain detailed diagnostic data from pets. But to what degree can

---

[32] Hypomorphic is a type of mutation in which the altered gene product possesses a reduced level of activity.

the mechanisms be comparable when using data from dogs to evaluate the likely impact of an environmental exposure on human populations? Must they be identical, or is there a way to normalize the concept of a canine sentinel based on genetic, histopathological, or other biomarkers that may not be the same as those in humans? When thinking about these questions from the standpoint of the exposure, said Rabinowitz, researchers are trying to determine what in the environment is carcinogenic. The traditional method looks at animal experimental studies, which often produce a different type of tumor than what's seen in humans, but nonetheless can reinforce the fact that a given substance or hazard is carcinogenic. The purpose is not to replace current methods but to introduce a whole new body of epidemiological evidence from companion animals to add to the epidemiological evidence from humans and the experimental evidence from animal exposure studies and in vitro work.

Walker commented that canaries did not go into coal mines because anyone thought they were little humans. It is useful to partition out hazard identification, exposure assessment, disease etiology/mechanism, and therapeutics. Each one of those endpoints would involve companion animals in a different way, said Walker.

METHODS AND CURRENT STUDIES

Speakers reviewed the current state of exposure science as it relates to methods, data, and analysis. Speaker presentations described examples of causal inference between exposures and cancer and aging, as well as potential data sources, including through the use of biosensors, needed to assess whether companion animals may serve as sentinels for human environmental exposures.

The Exposome and Health

The exposome idea, first introduced by Wild (2005), offers a way to consider the totality of an individual's environmental exposures over the course of a lifetime, said Ellison (see Figure 7). Wild defined the exposome as all exposures from conception onward, including those from lifestyle, diet, and the environment. Although this enables consideration of multiple exposures, it carries a significant challenge in terms of managing the data and integrating it into models of disease risk, said Ellison. Gary Miller (Columbia University) and Dean Jones from Emory University refined Wild's definition as "the cumulative measure of environmental influences and associated biological responses throughout the life span," emphasizing that the biological signal left by a stressful exposure, such as an epigenetic mark, may persist long after the initial exposure is no longer present (Miller and Jones, 2014). Miller stressed the need to measure external influences on health in an unbiased way to avoid

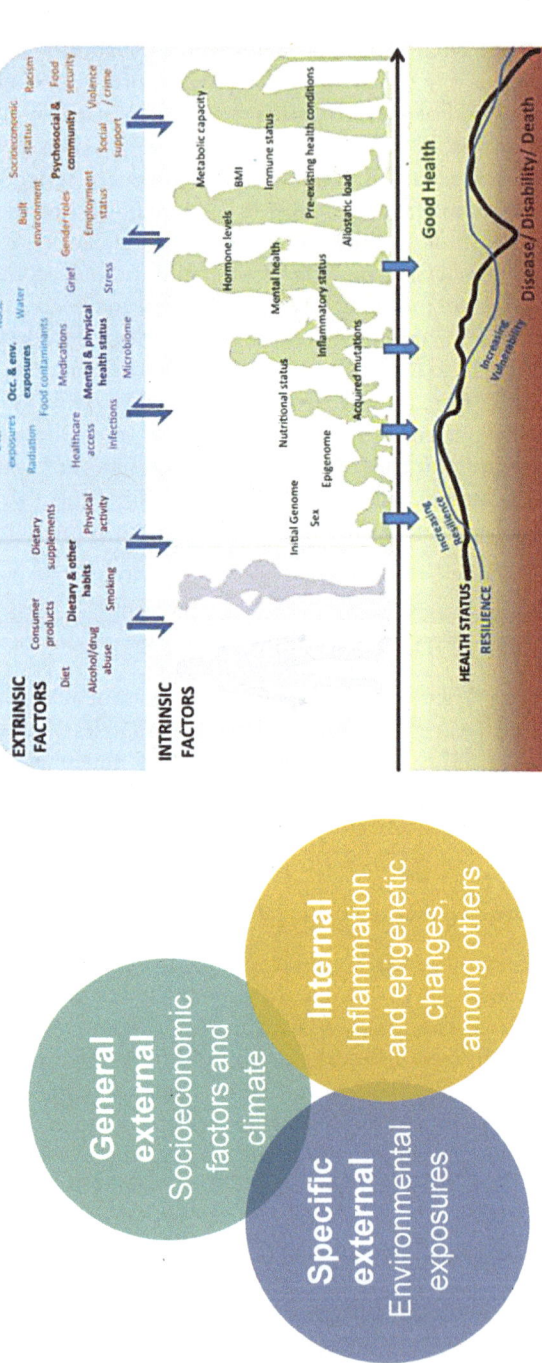

FIGURE 7 The exposome: Implications for examining environmental health disparities. External/extrinsic (*E*) and internal/intrinsic (*I*) factors interact throughout the life span, enhancing vulnerability or resilience in a cumulative manner. Many *E* factors act on and influence the *I* factors, whereas *E* factors are modulated to a lesser degree by *I* factors (indicated by thick and thin arrows between upper and middle panels). *I* X *E* interactions can vary over the life span, beginning before conception. In a given individual, *I* X *E* interactions influence vulnerability and resilience, and consequently health status, throughout life, as indicated by the fluctuating curves in the lower panel, which are hypothetical trajectories indicative of negative and positive effects on resilience and health status over the life span.
SOURCES: Ellison presentation, December 1, 2021. Illustration of intrinsic and extrinsic factors is reprinted from McHale et al., 2018.

making invalid associations and to create "a systematic, unbiased, and '-omic-scale examination of external factors contributing to disease or health status."

Detecting the Metabolic Signatures of Biological Exposures

A broad range of physical and chemical exposures affects both humans and their animal companions (Figure 8), but in order to have an impact on health, an exposure must be converted to a biological signal inside the body, said Miller. Capturing the breadth of these exposures and signals will require a suite of tools, from wearable devices and biospecimens to computational imaging and satellites. In his research, Miller pairs high-resolution mass spectrometry (HRMS) with both liquid and gas chromatography to detect a broad range of volatile and soluble chemicals in biospecimens. This effort is complicated by the limited availability of authentic standards for many exogenous chemicals and their metabolites, requiring Miller and colleagues to compare individual compounds' mass and retention times based on their peaks, which may be in the tens of thousands per measure, to PubChem and other databases containing nearly a hundred million compounds (Vermeulen et al., 2020). Miller added that this can be a "computational nightmare." Another challenge is the under-

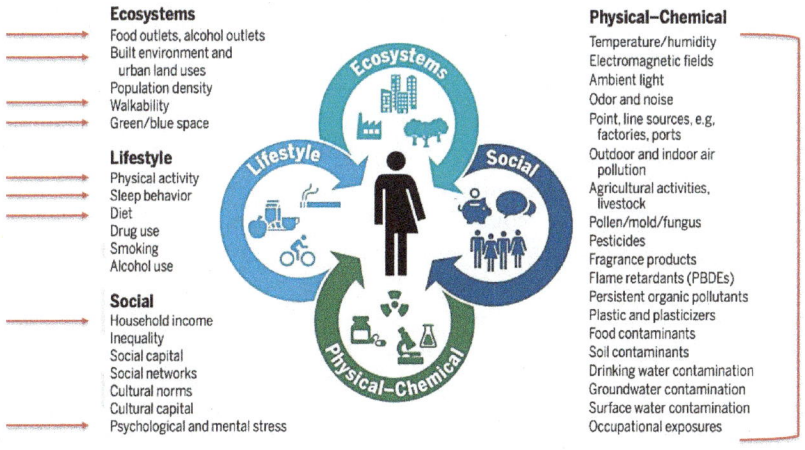

FIGURE 8 The exposome concept.
The exposome is an integrated function of exposures on the body over the life course. Factors shared among humans and their animal companions are marked with red lines. These include all the chemical stressors on the right side and many other types of stressors on the left.
SOURCES: Miller presentation, December 1, 2021. Adapted from Vermeulen et al., 2020.

representation of biological metabolites of exogenous chemicals in the databases; these metabolites may comprise the only biological signature of exposure to many compounds, so it is important to know what they are, said Miller.

Miller and colleagues designed a method for identifying the biological metabolites of a vast range of agents, in which they exposed human liver fractions to known environmental chemicals and characterized the products based on their peaks following mass spectrometry and chromatography (Liu et al., 2021). They validated this method by treating mice with one of the tested chemicals, buprorion, detecting the same profile of metabolites in mouse blood that they had previously identified in the liver assay. These same metabolites were also detected in a human taking buprorion pharmaceutically. The researchers were surprised to discover a novel metabolite of caffeine, suggesting the existence of a large number of unidentified metabolites overall. Miller acknowledged that many analytical chemists would bristle at the identification of a compound in the absence of an authentic standard, but he noted that standards only exist for a few thousand compounds, while his group analyzes a hundred thousand peaks, and matching the peaks to those obtained from the metabolites of known compounds is quite effective.

Applications of Exposomics to Human Disease

Miller explained that this approach is beginning to yield valuable information linking exposure to disease. An exposome-wide association study (EWAS) found an association between environmental chemicals in plasma and two rare liver diseases, while a metabolome-wide association study (MWAS) uncovered disease-associated alterations in metabolic pathways in the same cohort (Walker et al., 2021). Analysis of 119 individuals from Richard Mayeux's Washington Heights/Inwood Columbia Aging Project (WHICAP)[33] cohort distinguished between participants with and without Alzheimer's disease (AD), as well as among different racial and ethnic groups, based on metabolites in their plasma (Vardarajan et al., 2020). Miller noted that the degree of separation was unlike anything Mayeux had observed in this same cohort over the course of 30 years and that a follow-up study was being performed on 8,000 metabolomes to determine what metabolic pathways were involved. The researchers began a parallel study in the Dominican Republic, where many members of Mayeux's New York cohort originated. Miller noted that Mayeux has extensive environmental data from northern Manhattan for the past 50 years. Miller is also analyzing biobank plasma specimens from a cohort of individuals aged 20 to 80 years old who have been studied by cognitive testing and functional magnetic resonance imaging (fMRI) imaging (Stern et al., 2014).

---

[33] See https://dss.niagads.org/studies/sa000007 (accessed December 17, 2021).

Laboratory animal models may be used to test hypotheses generated by EWAS. For example, Miller and others described metabolites of dichloro-diphenyl-trichloroethane (DDT) that were elevated in humans with AD. Using an AD model strain of *C. elegans* containing a mutation in the tau gene, Kalia et al. (2021) found that tau mutant worms generated a distinct pattern of DDT metabolites compared to control worms. In a study of worms bearing a knockout of a Parkinson's disease–associated gene, changes in gene expression were correlated with patterns of metabolite production (Mor et al., 2020).

The disease-related metabolic similarities between species as evolutionarily distant as worms and people suggest that the potential for cross-species EWAS/MWAS analysis in companion dogs and humans is very strong, said Miller. A collaboration with Mark Prausnitz produced a microneedle patch that can sample interstitial fluid from skin without a blood draw, which may be helpful for sampling pets at home (Samant et al., 2020). Miller cautioned that developmental windows matter, citing a study on couples trying to conceive that found significant variability in their exposomes despite their shared living environment (Chung et al., 2018). This highlights the importance of knowing which stages of the human and animal life spans were co-exposed when doing paired cross-species studies, said Miller.

As exposomics gains acceptance in the biomedical research community, where it has been used to study diverse organisms and can be performed on blood samples from any species without technical limitations, it is important to know what you are trying to achieve, said Miller. He favors untargeted HRMS due to its relative lack of "streetlamp bias,"[34] and because so many metabolites have not been identified. Targeted approaches have greater sensitivity and confidence in identification, but they are entirely dependent on this tough question: "Do you know what you are looking for?"

Biomonitoring of Chemical Exposure in Companion Animals and Humans

Evidence of Exposure to a Wide Variety of Agents in Cats and Dogs

Kannan began the discussion of biomonitoring by reviewing the literature on exposure to heavy metals in pets. The handful of papers published on this topic demonstrate that cadmium, mercury, and lead exposure can cause neurodegeneration in pet cats and dogs, said Kannan. High levels of mercury

---

[34] Streetlamp bias, or streetlight bias, is when people search for something in easier places instead of in the places that are most likely to yield the results they are seeking; i.e., looking under the lamppost.

were found in the brain, liver, and kidney tissues of the Minamata "dancing" cats (Eto et al., 2001; see section on Evidence of Exposure to a Wide Variety of Agents in Cats and Dogs). Toxic levels of heavy metals have been found in pet blood, hair, urine, and organ tissue, and lead concentrations in dogs and in children from the same household and socioeconomic status show an association, said Kannan. The persistent organic pollutants polychlorinated biphenyls (PCBs) and DDT and their metabolites have been widely found in cat and dog tissues, with PCB levels linked to diabetes and hypothyroidism, said Kannan. California cats had high serum concentrations of the flame retardant PBDE (polybrominated diphethyl ethers), which is linked to hyperthyroidism and house dust exposures (Guo et al., 2012, 2016). Thyroid, kidney, liver, and respiratory diseases have also been associated with high serum concentrations of poly- and perfluoroalkyl substances (PFAS) in cats, whose patterns of exposure to these substances are similar to those observed in humans (Bost et al., 2016).

Kannan and colleagues detected higher levels of metabolites of phthalates—plasticizers—in urine from cats and dogs compared to those reported in humans, with cats having the highest levels overall (Karthikraj et al., 2019), and used reverse dosimetry to estimate the daily exposure dose. Shelter animals had lower exposure doses than house pets, suggesting that some exposure was occurring in the home. Testing for additional compounds, the researchers found that of the three species (cats, dogs, and humans), dogs had the highest levels of antimicrobial parabens in their urine, less than 10 percent of which was acquired from their diet; much higher levels of parabens were found in dry foods than wet foods, and more in cat food than in dog food (Karthikraj et al., 2018b). The group also detected the kidney toxicant melamine and the herbicide glyphosate in dog and cat urine (Karthikraj et al., 2018a; Karthikraj and Kannan, 2019) and PFAS in pet feces (Ma et al., 2020). Many urine samples from pets and humans had levels of melamine exceeding those associated with chronic toxicity, and levels in children were higher than those in adults (Karthikraj et al., 2018a). Calculated PFAS exposures were one to three orders of magnitude above the Agency for Toxic Substances and Disease Registry (ATSDR) provisional minimal risk levels (Ma et al., 2020). Microplastic levels in cat and dog feces were higher than those in adult humans but similar to levels in 1-year-old children (Zhang et al., 2019), likely because infants and pets chew on textiles and plastics, conferring much higher exposures than diet alone. These observations and others from his laboratory indicate that biomonitoring methods can assess exposures and risks from a wide variety of contaminants, said Kannan. While pets often show higher levels of exposure than humans, patterns of exposure are similar, particularly with respect to indoor environmental contaminants, and they do not primarily come from the diet. Kannan suggested a collaborative approach that combines clinical assessments of biomarkers in pets with paired human–pet biomonitoring studies to advance the use of pets as sentinels of environmental health.

Rabinowitz presented on unpublished research in which he and colleagues measured the levels of volatile organic compound (VOC) metabolites in dogs and humans living near natural gas wells, finding similar trends in exposure levels for both dogs and humans, with the dogs having higher levels of VOC metabolites than their human companions in 70 percent of cases, suggesting higher exposure in the dogs. As homes' distance from a gas well increased, there was a greater distance-related reduction in VOC metabolites in dogs than in humans, indicating that the metabolites constituted a likely exposure biomarker in dogs more than in humans. Nicole Deziel (Yale School of Public Health) noted that childhood cancers are regrettably being used as a "sentinel" for adult cancers in populations exposed to fracking, because of the shorter latency, and that both companion and domestic animals would be ripe for study. This argues for routinely including animals whenever you build a longitudinal cohort for observational studies of potential hazards, such as those related to natural gas drilling, said Rabinowitz.

Assessing the Exposome using Wearable Sensors

Environmental exposures show dramatic variation in type, timing, and dose across populations, time, and space, and therefore require a variety of tools for their detection and analysis, said Cui, and the goal of the NIEHS is to use these tools to understand how exposome–genome interactions influence health outcomes (see Figure 9).

The NIEHS sponsors the development and refinement of a variety of wearable and field-deployable tools for characterizing personal exposures to air pollution and for monitoring physiology. Although these are being developed for use in humans, they can be adapted for companion animals, Cui explained. The NIEHS-supported sensor research includes development of hardware, software, miniaturization, improvements to battery life, and laboratory and field testing to improve accuracy and reliability. Cui described five examples of this work (see Box 2).

Cui mentioned that there are challenges associated with analyzing large volumes of sensor data, as well as privacy concerns. But she noted that the number of PubMed publications on wearable sensors has been increasing exponentially in recent years, indicating the great potential of this field.

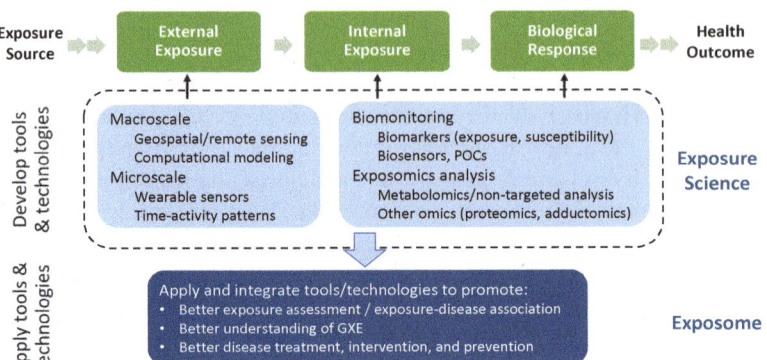

FIGURE 9 Approaches to exposure science at the NIEHS.
The NIEHS supports methodologies to address large-scale exposures, such as geographic information system mapping, as well as technologies to enable more precise estimates, such as personal monitors, and analytical methods to capture the comprehensiveness of the exposome, such as untargeted analysis and '-omics technologies, with the ultimate goal of applying these tools to better understand the exposome and its impact on health.
NOTE: GXE = gene by environment (gene environment interactions); POCs = point of care devices.
SOURCES: Cui presentation, December 1, 2021. Figure provided by the National Institute of Environmental Health Sciences.

Ongoing Canine Population Studies

The Domestic Dog as a Sentinel Species for Environmental Influence Associated with Bladder and Other Cancers

There are remarkable similarities between human and canine cancers, both genetically and from an anatomic pathology perspective, said Breen. Furthermore, health data and biological specimens from the 6 million dogs that are diagnosed with cancer each year provide an "enormous opportunity" for research, with millions of biopsy specimens from pet dogs stored around the United States. By cross-referencing data from dogs and humans for cancers that are considered comparable, Breen and colleagues have found many shared genetic features between human and dog tumors for a range of cancers (Angstadt et al., 2012; Mochizuki et al., 2015; Schiffman and Breen, 2015; Tawa et al., 2021). However, genetics are insufficient to explain why, among populations of dogs with the same genetic risk factors, some develop cancer and others do not. To understand how the genetics interact with environmen-

BOX 2
Examples of NIEHS-Sponsored Sensor Research

- The Biking & Breathing Study,[a] led by a group at Columbia University, used wearable sensors to study how air pollution from traffic affected the health of New York City cyclists. The hypothesis was that potential inhaled dose, calculated by multiplying the concentration of air pollution by the minutes of respiration, was a better metric to study the impacts of air pollution on heart rate and blood pressure than data from fixed air quality monitoring stations. Wearable monitors were used by 149 cyclists over five-to-six 24-hour sessions. Physiological sensors recorded cyclists' respiration, blood pressure, and heart rate while an accelerometer recorded physical activity, a GPS mapped locations, and air monitors measured small particles (particulate matter smaller than 2.5 um [PM2.5]) and black carbon, enabling a determination of individual exposures throughout the city in real time.
- The company MyExposome, Inc.[b] is developing silicone wristbands for use as personal samplers, which is supported by the NIEHS through the Small Business Program. By using paired bands, with-versus-without skin contact, researchers are building a novel predictive model of exposure that can detect 1,500 chemicals, with the aim of distinguishing those that come from the air alone.
- For continuous monitoring of internal biomarkers, a group at the University of New Mexico Health Center[c] is developing minimally invasive wearable microneedle arrays that sample the interstitial fluid for several hours, after which the arrays can be removed and analyzed for exposure to heavy metals. The goal of this project is to overcome a major challenge to developing microneedle array biosensors, which is the extremely low abundance of exposure biomarkers in the interstitial fluid compared to endogenous molecules (García-Guzmán et al., 2021; Rappaport et al., 2014).
- Perry Hystad and colleagues at Oregon State University are developing an algorithm to leverage self-tracking cell phone GPS data to link environmental exposures to a person's physical activity and the built environment. The researchers obtained historical cell phone location data from roughly 40 individuals that mapped each person's activity for over a thousand days.

continued

> **BOX 2 Continued**
>
> - The All of Us Fitbit longitudinal study[d] leverages self-tracking wearable devices to collect physical, physiological, and tracking data from over 10,000 participants that can then be combined with environmental and health data.
>
> SOURCE: Cui presentation, December 1, 2021.
>   [a] See https://asic2018.aqrc.ucdavis.edu/sites/g/files/dgvnsk3466/files/inline-files/Darby%20Jack%20-%20Jack_ASIC_september2018_oakland%20%281%29.pdf (accessed December 20, 2021).
>   [b] See http://www.myexposome.com (accessed February 7, 2021).
>   [c] See https://tools.niehs.nih.gov/portfolio/index.cfm/portfolio/grantDetail/grant_number/R03ES031724 (accessed December 20, 2021).
>   [d] See https://www.joinallofus.org/go (accessed February 7, 2021).

tal risks to produce cancer, "we need to understand a lot more about our dogs' exposures, to understand more about our own exposures," said Breen. To do this work, Breen and colleagues built a nationwide collaboration among dog owners, scientists, veterinarians, and breeders, with two fundamental goals: (1) to monitor changes in cancer incidence in dogs in different geographical regions, using the shorter latency of dog cancer to identify risks to humans, and (2) to engage dog owners in clinical trials that include exposure assessment. The best strategy, says Breen, is to do paired exposure studies of humans and dogs from the same household, where one of those species has been diagnosed with cancer.

In a collaboration between North Carolina State Veterinary School and Duke University, Breen and colleagues have begun enrolling dog patients in combined clinical–environmental studies, the first of which is looking for environmental associations with bladder cancer in dogs, who are typically diagnosed at age six or older. The researchers developed a 'liquid biopsy' that enabled them to detect very early bladder cancer based on gene signatures in the urine, allowing them to "wind the genomic clock back" and look for the earliest genetic changes. Working with the American Kennel Club (AKC), they recruited owners of 2,000 dogs aged 6–12 years from seven breeds with a high predisposition for bladder cancer. Their first project, involving use of silicone wrist bands and matched urine to investigate shared exposures, has revealed "striking . . . similarities between exposures of our dogs and ourselves," said Breen (see section on Using Silicone Samplers to Assess Air Pollution in People and Pet Dogs).

Breen noted that the number one risk factor for bladder cancer in humans is smoking, but unlike lung cancer, bladder cancer has not declined with the reduction in smoking over the last 30 years. This suggests that other environmental factors may be at play; for example, certain pockets of Ohio are bladder cancer hot spots. In a collaboration with the Cleveland Clinic, Breen's group is enrolling human patients with bladder cancer and their dogs in a study to monitor exposures through water samples, urine, and wrist bands.

Bladder cancer differs between dogs and people and this complicates comparisons, suggested Beverley Koller (University of North Carolina). Eighty to 85 percent of dog urothelial carcinomas (bladder cancer) has the equivalent of a human V600E BRAF[35] mutation, which is present in only 1 to 2 percent of human bladder cancers, agreed Breen. However, he noted that there is regional variation in human bladder cancer, with the V600E mutation present in as many as 25 percent of human patients in certain cohorts. Mutational signatures need to be considered in terms of their geospatial distribution in order to explore the causes of regional variation, even within a state, said Breen.

A third study builds on an existing North Carolina cohort from the NIEHS Personalized Environment and Genes Study (PEGS).[36] Participants have contributed genotype, health, and environmental exposure data since 2002. Breen aims to obtain genetic and health information from participants' dogs, overlay this onto the exposure maps that were developed for the human study, and determine whether exposures affect the health of dogs before that of humans. Lastly, as part of a large collaboration with Timothy Mousseau (University of South Carolina), Ostrander, Norman Kleiman (Columbia University), and others, Breen is studying the dog populations that have roamed the area surrounding the Chernobyl nuclear plant since the reactor melted down in 1986, as a model for the generational health impacts of chronic exposure to radiation, heavy metals, and other environmental toxins.

The Golden Retriever Lifetime Study

Rodney Page (Colorado State University) described the Golden Retriever Lifetime Study (GRLS), which emerged from an effort of the Morris Animal Foundation to make progress on canine cancer prevention that started 15 years ago (Box 3).

Participants were interested in the extent to which the longitudinal dog studies investigated drinking water. The GRLS surveys owners about water

---

[35] BRAF is a gene found on chromosome seven that encodes a protein also called BRAF. This protein plays a role in cell growth by sending signals inside the cell promoting, among other functions, cell division. The V600E BRAF mutation makes a protein that is involved in sending signals in cells and in cell growth

[36] See https://www.niehs.nih.gov/research/clinical/studies/pegs/index.cfm (accessed February 7, 2022).

> **BOX 3**
> **The Golden Retriever Lifetime Study**
>
> The Golden Retriever Lifetime Study (GRLS)[a] is an observational study that seeks to produce valuable information on canine cancer, including incidence and risk factors, both genetic and environmental (Guy et al., 2015). Additional goals include developing new early detection technologies, biomarkers, and prevention strategies, and follow-up studies may eventually test interventions. The GRLS is also an aging and longevity research study, collecting data and specimens in a systematic way from cradle to grave, which could be used to study a variety of questions other than those related to cancer, said Page.
>
> Golden retrievers are popular pure-bred dogs with a high risk of developing and dying from cancer (Guy et al., 2015). Following a successful recruitment campaign using social media, a total of 3,040 dogs were enrolled from across the United States between 2012 and 2015, with 60 percent suburban, 30 percent rural, and 10 percent urban. Both male and female genders, overweight/obese dogs, and intact/neutered animals are well represented in the study population. Enrollment required consent of owners, who had to agree to care for their dogs for the lifetime of the dog, and of veterinarians, who had to agree to complete the requirements of the study. Diagnosis of one of the four types of cancer most common in golden retrievers was chosen as a primary endpoint, with an aggregate of 500 of these primary endpoints forming the study endpoint.
>
> Annual data collection includes questionnaires for both owner and veterinarian as well as collection of biospecimens. The questionnaires follow the general concept of the Nurses' Health Study[b] and ask about pedigree, including sire, dam, and littermates; reproductive history; behavioral, social, physical, and cognitive function; and exposures, including insecticides, herbicides, indoor and outdoor air quality, diet, water, and lifestyle. When possible, tumors that arise undergo histopathologic analysis by two or more pathologists. The researchers developed a strategy for adjudication of tumors that enables both tumor identification and degree of certainty of the diagnosis, which is incorporated into

source (bottled, tap, well) as well as filtration, type of pipes, and regional ground water contamination, said Page. Lauren Trepanier (University of Wisconsin) cautioned that handling of the water significantly impacts the ability to detect different chemicals. In two prospective studies, Trepanier explained the data. To facilitate retention, an online support network was formed for owners.

In June 2021, the median age of the cohort was 8.5 years and of the 3,040 dogs enrolled, 352 had died, 66 percent due to cancer. The study endpoint was expected to be reached in 2023, said Page. The biorepository contained over a million stored specimens of serum, blood, urine, hair, toenails, feces, tumor, and normal tissues.

Page noted that 20 percent of the dogs were reproductively intact, and for the others, the age at spay/neuter was an important consideration, as those neutered early never went through puberty, and later spay/neuter has been suggested to correlate with the emergence of certain risk factors. Over 1,000 puppies had been born, adding an F1 generation to the cohort.

Of the cancers diagnosed by June 2021, hemangiosarcoma was the most common, representing 50 percent of all tumors, followed by lymphoma, histiocytic sarcoma (which was not included in the original four), mast cell tumor, and osteosarcoma. Ten years into the study, compliance with the annual data collection by owners and veterinarians was 85 percent.

Additional questions that can be addressed include studies of obesity, exposure to pesticides and herbicides, air quality, and water; "it goes on and on . . . in terms of creating hypotheses for investigation," said Page. A control group of aging (>12 years) golden retrievers that did not have cancer was currently being added and undergoing genotyping in the "Golden Oldies Project." This cohort, while strong, was very homogenous, which may limit some of its representativeness, said Page.

The GRLS Data Commons[c] was developed to enable investigators around the world to access data and specimens. It includes both cancer- and non-cancer-related conditions. Researchers can submit proposals to access biospecimens and data from the questionnaires.

SOURCE: Page presentation, December 1, 2021.
  [a] See https://www.morrisanimalfoundation.org/golden-retriever-lifetime-study (accessed February 7, 2022).
  [b] See https://nurseshealthstudy.org (accessed March 30, 2022).
  [c] See https://www.morrisanimalfoundation.org/data-commons (accessed February 7, 2022).

that she uses four different collection methods to test for five chemicals in the drinking water of dogs and their owners.

One participant remarked that the size of the GRLS could make it possible to identify genetic risk factors for cancers that occur less frequently in the golden retriever but are more common in other breeds, such as histiocytic sarcoma. Page noted that it was too early to associate risk factors with tumors, but the ability to compare littermates and the availability of over 1,000 puppies could enable a study of the intergenerational effects of exposure, cost permitting.

The Dog Aging Project: A One Health Perspective on Healthy Aging

Daniel Promislow (University of Washington) described the Dog Aging Project (DAP) (Box 4), which was initiated 14 years ago as a collaboration with Kate Creevy at the University of Georgia, inspired by Ostrander's work

---

BOX 4
The Dog Aging Project

The Dog Aging Project (DAP)[a] is a long-term longitudinal study focused on the biological and environmental determinants of healthy aging, said Promislow. Owners nominate their dogs for enrollment through a form on the project website, and all dogs in the United States are welcome to join regardless of breed, age, sex, size, reproductive status (intact/sterilized), or state of residence. Ultimately, this study aims to enroll 100,000 dogs in the United States, and at some future time, to expand internationally.

Since the website was launched 2 years ago, the owners of more than 37,000 dogs had completed the Health and Life Experience Survey, and over 18,000 of those had uploaded electronic medical records, making them eligible to join a sample cohort. Dogs are randomly assigned to one of three cohorts:

- *Foundation cohort*—8,500 dogs were assigned to this cohort, with owners receiving a cheek-swab kit in the mail, which is analyzed by low-pass whole-genome sequencing. Three thousand dogs had been genotyped to date, and Promislow expected to have 10,000 genotypes by the summer of 2022.
- *Precision cohort*—Owners of 1,000 dogs receive larger kits that the veterinarian uses to collect whole blood, serum, plasma, urine, feces, and hair specimens, which are shipped to the diagnostics lab at Texas A&M Veterinary Medical Diagnostic Laboratory (TVMDL) for standard clinical chemistry. The TVMDL distributes specimens to multiple labs for additional analysis. Residual specimens are biobanked in the DAP biobank at the Cornell University College

and Page's Golden Retriever study, and now involves a team of about 100 scientists in addition to many thousands of dog owners.

The DAP can be understood from a quantitative genetics perspective, said Promislow, where G = variation in genes, E = variation in the environment, P = variation in phenotype, and $G + E \rightarrow P$. For genetic information, DAP uses whole-genome sequencing or owner-provided breed identification, which is less reliable. Environmental information is derived from both owner-completed surveys and other sources pegged to zip codes. Phenotypic information includes behavior, age-specific disease, frailty, comorbidities, mortality, and fecundity. To probe the intermediate mechanisms by which genes and environment trans-

---

of Veterinary Medicine for use by scientists worldwide. There were plans for the precision cohort to receive activity monitors.

- *Intervention cohort*—By the middle of 2023, Promislow expects the DAP to have enrolled about 500 large-breed, middle-aged dogs in a double-blind placebo-controlled study of rapamycin. The primary endpoint is survival and secondary endpoints are measures of heart and cognitive function. Biospecimens are collected, processed, analyzed, and biobanked as for the Precision cohort.

To generate longitudinal data capable of hypothesis testing, owners complete annual online surveys. In addition to the health and life experience survey, these include behavioral, cognitive, environmental, and end-of-life surveys. Data are returned to the dog owners; this form of reciprocity is essential for citizen science, said Promislow.

Data are embargoed for 1 year for analysis by DAP trainees, after which they are released publicly on the project website.[b] Release 1.0 will become publicly available in the Spring of 2022, and Release 2.0 (the second year of data) will come out at the same time for use by the internal team. Summary data are available, and external researchers can access embargoed data through collaborations.

Diversity is a concern for the DAP, with 90 percent of enrolled owners self-identifying as white non-Hispanic and tending to be wealthier and more educated than average, and fewer than 1 percent Black non-Hispanic. Promislow explained that he is exploring ways to increase diversity, such as elimination of the electronic medical records requirement. Dog demographics are split evenly by gender and by purebred vs. mixed breed, with over 90 percent spayed or neutered.

SOURCE: Promislow presentation, December 1, 2021.
[a] See www.dogagingproject.org (accessed December 28, 2021).
[b] See http://data.dogagingproject.org (accessed February 7, 2022).

late to phenotype, the DAP uses tools of systems biology to analyze clinical chemistry, metabolome, epigenome, microbiome, and transcriptome, as well as fluorescence-activated cell sorting, in the 1,000-dog Precision cohort.

Scientific questions being addressed by the DAP are split into four projects: (1) Aging, to create a measure of normative aging in dogs that can inform the care of dogs as they age; (2) Genetics of Aging, to understand the genetics of aging as well as gene–environment interactions in dogs; (3) Systems Biology of Aging, to understand the '-omics of aging in dogs; and (4) TRIAD (Test of Rapamycin in Aging Dogs), the clinical trial. Other clinical trials, including tests of cancer treatments, can be envisioned in the future, said Promislow.

Promislow described several analyses that the DAP consortium planned to submit for publication in the coming months: three different measures of activity mapped a steady decline in activity with age of the dog; rural dogs were more active and for more time than urban and suburban dogs; and older owners rated their dogs' activity levels differently than younger owners. In an analysis of meal frequency and health, eating just one meal daily, which is a type of intermittent fasting (Mattson et al., 2017), correlated with a lower rate of various diseases in dogs. Promislow noted that this in itself does not imply causation, as sicker dogs may eat more often, but diversity of eating patterns offers an opportunity to test the health effects of various forms of diet restriction on aging. The DAP collected a great deal of data on the frequency and types of cancers in various dog breeds and on the environment. Older owners rated dogs differently than younger owners in several areas, suggesting that owners' answers to survey questions "will partly be a reflection of themselves and not their dogs," said Promislow.

The DAP is an open science project, said Promislow; all data will be made available through the Terra platform at the Broad Institute of MIT and Harvard, which acts as a wrapper around the Google cloud, facilitating analysis or downloading of data. Promislow encouraged scientists to propose ancillary studies that can build on the DAP infrastructure and generate future avenues for research, making DAP a "forever study."

Companion Dogs as Sentinels for Chemical Effects on Male Reproductive Function

Richard Lea (University of Nottingham) discussed research using companion dogs as sentinels for environmental effects on male reproductive function (Box 5).

One environmental factor that varies among countries is the available brands of dog food, and these differences can influence the microbiome, noted

Andrei Gudkov (Vaika Inc. and Roswell Park Cancer Institute). To determine exposures from food, Lea analyzed chemicals in a dozen different popular types of dog food in the United Kingdom; he found that the same chemicals elevated in semen and testes were also elevated in certain brands of food, suggesting that food is a major cause of exposure of these contaminants. While his hypothesis was that exposure to these chemicals during the developmental period was responsible for the observed symptoms of testicular dysgenesis syndrome (TDS), the fact that they have short-term effects on sperm motility in vitro suggests a potential effect at later life stages as well, regardless of exposure in utero, Lea added.

BOX 5
Studies of Dogs as Sentinels for
Testicular Dysgenesis Syndrome

Recent decades have seen declining sperm counts in men in North America, Europe, and Australia (Levine et al., 2017), along with increasing rates of testicular cancer (Reese et al., 2021), cryptorchidism (lack of testicular descent in fetal development) (Le Moal et al., 2021), and hypospadias (urethra opening toward the base of the penis) (Xing and Bai, 2018). These four problems may occur simultaneously, suggesting a common etiology, and they are referred to collectively as testicular dysgenesis syndrome (TDS). Incidence of TDS shows geographic clustering—for example, rates of all four disorders are higher in Denmark than Finland—and that has implicated the environment as a potential cause (Xing and Bai, 2018).

Noting that the early stages of development, when reproductive organs are forming, are exquisitely sensitive to environmental chemicals, Lea sought to determine whether reproductive health is declining in companion dogs by looking at temporal trends in sperm quality, testicular function, testicular cancer, and malformations in pups.

Study 1: Fertility in a population of stud dogs
Lea and colleagues studied a population of 232 healthy stud dogs from five breeds (Labrador retriever, golden retriever, curly coat retriever, border collie, and German shepherd), examining a total of 1,925 ejaculates collected during routine exams from 1988 to 2014. Over this 26-year period, the researchers found a 30 percent decline in sperm motility, paralleling the decline observed in

continued

BOX 5 Continued

humans during the same period, and an increase in cryptorchidism in the pups (Lea et al., 2016). Chemical toxicants, including congeners of PBDEs (flame retardant) and PCBs (industrial coolant and plasticizer), were detected in semen, milk, ovaries, and testes of dogs from the same population as the original animals (Lea et al., 2016). When they incubated dog sperm with two of these chemicals in vitro, the researchers found that di(2-ethylhexyl)phthalate (DEHP) and PCB 153 caused DNA fragmentation in the sperm at concentrations similar to those detected in dog gonads (Sumner et al., 2019). Both chemicals showed similar dose-dependent effects on human and dog sperm, supporting the hypothesis that the dog can serve as a sentinel for chemical exposures associated with TDS, said Lea.

Study 2: Testicular cancer in humans and dogs

The incidence of testicular cancer in humans shows regional and temporal variation, which suggests environmental causality, but no specific cause and effect has been conclusively demonstrated, said Lea. There has been a simultaneous increase in testicular cancer in dogs (Grieco et al., 2008a) as well as detection of cancer precursors in dog testes (Grieco et al., 2008b). To interrogate the association of environmental factors with testicular cancer in dogs, Lea and colleagues studied testes collected from castrated dogs from Finland, Denmark, and three regions within the United Kingdom. The researchers established histological parameters for abnormal appearance of the testes and scored them based on the percent of each specimen that exhibited these abnormalities. They also profiled the testes for chemical pollutants.

The researchers saw a geographical difference in testicular pathology, with all three UK regions producing higher (worse) pathology scores than Scandinavia, and with Denmark worse than Finland (similar to human TDS), suggesting an environmental effect. Testes of Finnish dogs also had the lowest level of cancer precursors. Interestingly, while Finnish testes had the lowest levels of DEHP and PCBs, they had the highest concentration of PBDEs, followed by Denmark. Lea observed that, given the similarity between humans and dogs with respect to temporal changes in TDS, geographical variation of TDS, and chemical in vitro effects on sperm, dogs can help establish exposure-effect causality of TDS. Additional dog data on reproductive history, diet, and chemical exposures obtained through wearable monitors would be particularly valuable, said Lea.

SOURCE: Lea presentation, December 1, 2021.

Discussion: Advancing Use of Companion Dogs as Sentinels

Addressing Causality and Establishing Best Practices

Participants discussed the difficulty in establishing causality of environmental exposures on health outcomes. This would be difficult using dogs alone, said Lea, who suggested a combination of approaches in multiple species. For example, once similar associations are identified in parallel human and dog studies, a rodent model could be used to test controlled exposures, singly or in combination. Another way to address causality is through highly exposed populations like those in Minamata, offered Kannan, who suggested identifying highly exposed communities and then looking for cancers and other diseases in the pets, followed by laboratory exposure studies in rodents to confirm causality. For instance, if a high incidence of cancers were found in pets in a given geographic region or location, studies focusing on etiology would help strengthen cause–effect linkages. Miller suggested drawing on human examples that have strong causation, like the effect of cigarette smoking on cancer or of air pollution on pulmonary outcomes (e.g., emergency room visits for asthma), and extending the data collection to include an assessment of dogs in the same environment. Jan Dye (EPA) considered the GRLS in the framework of DOHAD, the Developmental Origins of Health and Disease.[37] The ability to compare exposures in breeding females and monitor their pups prospectively raises the potential of studying the effects of gestational and early life exposures on disease, or even just to measure the kinetics of exposure to determine the extent of placental transmission of agents or the duration of their transfer through the milk, said Dye.

Breen reflected on instances where purebred littermates living in the same household develop the same cancer within a short time window, while their littermates in different households are unaffected. Once a region with a high incidence of cancer in a particular breed has been identified, littermates living in regions with no similar environmental exposure can serve as controls. The result is "almost like a twinning study," said Breen, with sextuplets or octuplets, albeit not identical twins. This is where dog researchers will get pushback from the human research community because of "statistics, statistics, statistics," said Ostrander. "Any scenario . . . that sounds fascinating, the first thing they're going to say is, well, so how many? How many families of how many individuals into how many urban zones defined by zip code with what demographic, and then did you measure environmental exposures . . . exactly the way we want . . . and was it done the same all the way across the country?" In order to elevate the dog model as a sentinel, researchers need to design experiments together with bioinformaticians, genetic epidemiologists, and statistical geneticists "and figure out how to do this right," she added.

---

[37] See https://dohadsoc.org (accessed January 13, 2022).

Unlike humans, only about 3 percent of pets have health insurance, which influences the choice of treatment, said Chad Johannes (Iowa State University). Participants noted that this also limits the amount of data that can be obtained. The number of dogs with cancer is underestimated because many never receive a definitive diagnosis or even see the vet because pet owners cannot afford the tests and treatments.

One participant suggested adding environmental screening diagnostics for pets to regular periodic veterinary care, for the purpose of a greater public good. This could be useful for human patients as well, replied Myrtle Davis (Bristol Myers Squibb) and Miller.

Participants also considered how to develop best practices for veterinary research. Promislow commented that historically there has not been much support for large-scale veterinary studies, and the veterinary community needs to have a seat at the table to ensure that researchers are asking the right questions.

Systematic Tracking of Environmental Exposures in Dogs

Participants considered current efforts to systematically track environmental exposures in canines. This would require a definition of the normal distribution of chemicals in dogs—a reference exposome that could be developed from the current populations in biobanks, said Miller. Promislow, who is asking targeted questions focused on a limited set of exposures in the DAP, noted that a wholesale exposomic study of his specimens would be ideal, albeit cost prohibitive.

The National Health and Nutrition Examination Survey (NHANES)[38] is a decades-long reference study run by the Centers for Disease Control and Prevention (CDC) that measures several hundred different chemicals in a representative population of humans across the United States, but it is a series of snapshots of different people over time, not a longitudinal study of the same people. A study that follows dogs and people in the same homes over time would be extremely valuable, said Promislow.

The need for standardization of the exposome platform was mentioned by several participants. Chemical analysis is currently performed on multiple platforms in different labs using different extraction protocols without a reference set. The only solution to reduce variability, said Miller, is for multiple labs to use the same equipment. Miller suggested using HHEAR network labs, which are already standardized, for exposome analysis of large canine studies. But Breen said that labs can harmonize their approaches and cross-reference data without using the same equipment and protocols. Miller responded that this

---

[38] See https://www.cdc.gov/nchs/nhanes/index.htm (accessed January 18, 2022).

works well for targeted searches, where national and international standards already exist, but not for untargeted searches.

Aging Dogs as a Model for Cognitive Dysfunction

The utility of dogs as a model for human cognitive disorders was briefly discussed. A DAP study recently demonstrated a positive correlation between amyloid beta levels in dog brains and their cognitive dysfunction scores prior to death (Urfer et al., 2021). Promislow noted that while large-breed dogs are on a quicker trajectory than small breeds regarding life span and cancer incidence, all breeds seem to have a similar age trajectory for cognitive dysfunction, which typically begins after age 12, so the smaller breeds tend to display it more because they live longer (Watowich et al., 2020). Miller said that he preferred the term cognitive decline, which can be measured across multiple species, rather than dysfunction. He noted that recent improvements in biomarkers, like phosphorylated tau protein, have made it more feasible to measure correlates of pathology in dogs. Finch added that, unlike rodents, dogs naturally accumulate plaques and tangles with aging, produce the same amyloid peptide as humans, and have the same distribution of Alzheimer's-like pathology in the brain. Geriatrics is an important specialty that is currently missing from board certification in veterinary care, said Promislow.

Use of Electronic Medical Records in Research: Opportunities and Limitations

There is a need for standardized data models in canine population studies, said Anne Thessen (University of Colorado). Page and Promislow noted that electronic medical records (EMRs) systems in veterinary medicine are extremely variable and proprietary, and a useful standard for answering scientific questions would likely require a new build with standards compatible with those used for humans. The Integrated Canine Data Commons at the NIH may prove a useful foundation, said Page.

One goal of this research is to overlay human and canine electronic medical records (EMRs) with environmental exposure maps and geo databases, such as those being developed by the NIEHS in North Carolina, said Breen and DeGregori. This is expensive and requires coordination of human clinicians, veterinary medical professionals, and data scientists, noted Breen.

Raj Boya (private sector IT developer) asked whether current IT applications are meeting researchers' needs for data collection, storage, retrieval, and model design. The DAP researchers are harmonizing a variety of data types into a single resource that can be efficiently stored and accessed, offered Promislow. In addition, the DAP has responded to an NIH request for proposals to improve the readiness of existing large, complex data sets for analysis by artificial intelligence and machine learning (AI/ML). Page and Promislow noted that it is much easier to harmonize dog and human '-omic data than clinical information. The first challenge, said Promislow, is to harmonize the data within a study, which may include thousands of variables. For the sake of reproducibility, it is important to document the methods of data collection—the metadata, said Cui.

RELEVANCE OF COMPANION ANIMAL EXPOSURES TO HUMAN CANCER AND AGING

Participants discussed the science connecting specific exposures with cancer and aging outcomes in companion animals. Speaker presentations considered the state of the science of environmental exposure monitoring in companion animal populations, the benefits and opportunities in conducting biomonitoring in companion animals for cancer and aging research, how companion animal data can be leveraged to inform human monitoring, and the challenges and barriers to implementing exposure monitoring studies in companion animals.

Exposures to Air Pollution, Smoking, and Lead

Ambient air pollution and cigarette smoking are responsible for the deaths of over 16 million people each year globally, said Finch (Trumble and Finch, 2019). This is twice the number of pre-COVID deaths from infectious disease and 10 times those from road injuries (WHO, 2018), and this loss of life expectancy is getting worse due to industrial use of fossil fuels. Roughly one-fifth to one-third of dementia in humans can also be attributed to air pollution or smoking (Barnes and Yaffe, 2011; Cacciottolo et al., 2017; Chen et al., 2017). Tobacco smoke and air pollution have similar effects on both carotid artery disease and cognitive decline in adults (Baldassarre, 2009; Gatto et al., 2014; Künzli et al., 2010), and the two interact synergistically to increase the risks of lung cancer, brain aging, metabolic disorder, and atherosclerosis (Forman and Finch, 2018).

Dogs and Cats as Sentinels for Air Pollution

There's no question that dogs are a sentinel for air pollution, said Finch, citing postmortem studies that compared dogs in Mexico City, where the air contains high levels of particulates and polycyclic aromatic hydrocarbons, to those in coastal cities with far less pollution (Calderón-Garcidueñas et al., 2001a, 2001b; Calderón-Garcidueñas and Duyckaerts, 2017). The Mexico

City dogs had significant neuron loss in one region of the brain stem and their myocardium contained silica-like crystals. Two studies examined particulate exposures in Alaskan sled dogs while also testing new therapeutic approaches to mitigate the negative impacts of exposure, said Finch (Montrose et al., 2015; Witkop et al., 2021). Another study of secondhand smoke also provides an example of the utility of cats as a potential sentinel for young children, said Dye. While exposure to secondhand smoke (SHS) has been shown to cause lung cancer in humans, it may also put children at increased risk for lymphoma and leukemia (Rial-Berriel et al., 2020). Cats, like children, experience an increased risk of malignant lymphoma associated with secondhand smoke (Bertone et al., 2002), and this may be due to an additional exposure through their grooming, said Dye. In contrast, medium- or short-nosed dogs are at increased risk for lung cancer from secondhand smoke, while long-nosed breeds are at increased risk for nasal cancer (Reif et al., 1992, 1998).

Finch and colleagues proposed the concept of the "gero-exposome" to describe gene–environment interactions that produce aging phenotypes (Finch and Haghani, 2021). Many components of these exposomes apply equally as well to domestic animals as humans, he said. Finch raised the possibility of a connection between secondhand smoke and dementia in humans, which could be investigated in companion dogs because canine brain amyloid Aβ42, which correlates with cognitive dysfunction, increases with aging (Urfer et al., 2021), He also noted that exposure to air pollution increases amyloid production in mice containing the ApoE4 risk allele for Alzheimer's disease (Cacciottolo et al., 2017). Age, sex, and ApoE alleles all modulate the effect of pollution on the expression of a cluster of genes in mice that contribute to amyloid production (Haghani et al., 2020), providing a molecular basis for studying this effect. In the same study, mice that had been exposed to inhaled air pollution during gestation exhibited deficits in adult neurogenesis, behaviors, and fat metabolism that differed by sex and ApoE allele, with increased adult obesity and glucose intolerance.

Pets as Sentinels for Exposures Related to Low SES

The socioeconomic gradient has a huge influence on life span in humans, essentially shifting the entire survival curve by 15 years (Crimmins et al., 2009). This effect is larger than any of the genetic influences known to impact aging and is caused entirely by environmental factors, said Finch, with "not a shred of serious evidence that there are genetic differences [at play] except as gene–environment interactions." Low socioeconomic status (SES) is associated with increased exposures to life-shortening environmental factors, including cigarette smoking, which exposes more than 35 percent of children in the United States to secondhand smoke during their development (Tsai et al., 2018). Low SES also impacts animal welfare, with significantly more pets surrendered to shelters in areas with the highest poverty (Dyer and Milot, 2019; Ly et al., 2021; Monsalve et al., 2018).

Finch said that emerging literature is associating impulsivity in children with lead exposure, and there may be related animal behaviors, though they have not been examined (Dignam et al., 2019). Use of leaded gasoline caused an "immense public exposure, which is still being worked out" in studies on children's development, he added, noting that stray dogs and cats in Naples, Italy, had high levels of lead and cadmium in their livers and kidneys (Esposito et al., 2019). Dogs in Flint, Michigan, however, showed a low rate of elevated blood-lead levels with no clinical-grade lead toxicity (Langlois et al., 2017), though behavioral impacts were uninvestigated. Paint is now responsible for the majority of children's blood-lead elevation and there are very few case reports of lead-paint toxicity in pets, though it merits further investigation, said Finch.

Using Silicone Samplers to Assess Air Pollution in People and Pets

On average, people spend more than 90 percent of their time indoors, where they are exposed to a number of different chemicals, particularly organic contaminants, at higher levels than from exposure to chemicals outdoors, said Heather Stapleton (Duke University). Building materials and common household items that are chemically treated slowly emit these chemicals into the air over time, and they can accumulate indoors. Examples of such chemicals include flame retardants, synthetic dyes, and stain repellants in furniture and carpets; plasticizers, metals, and flame retardants in electronics; as well as components of insulation, drywall, and flooring. Thousands of different chemicals are detectable in indoor dust, though studies typically focus on only a dozen to a few hundred compounds.

Silicone Samplers for People

Contaminant exposures are typically measured by detection of chemicals in tissues such as blood, urine, and hair. Chemicals with half-lives of months to years are usually measured in blood, and those with shorter half-lives in urine. Industry has moved to using compounds with short half-lives, to reduce the use of persistent chemicals, which is leading to a greater reliance on urine to support biomonitoring studies (except for PFAS), said Stapleton. However, urine has several drawbacks. Metabolites measured in urine may be derived from multiple chemicals, making it difficult to trace a metabolite back to its parent chemical; and the concentrations of individual compounds in urine can show dramatic variability within a day and from one day to the next. To overcome these limitations and get a fuller accounting of the exposome, which requires an assessment of cumulative exposures over time, Stapleton and colleagues have been testing wearable samplers in the form of silicone wristbands. These wristbands are made of polydimethylsiloxane (PDMS), which absorbs chemicals well and can be used to measure ambient exposure to a variety of compounds (O'Connell et al., 2014), particularly those that are inhaled or absorbed through the skin. Chemicals in the gas or particle phase adhere to the silicone well and desorb poorly, even in the shower, so that all but the most volatile organic compounds accumulate over time. Exhaustive solvent extractions are used to pull compounds off the samplers for analysis.

A pilot study with 40 participants found that the amount of a common flame retardant, TDCIPP,[39] collected from silicone wristbands worn for 5 days correlated well with the amount of a TDCIPP metabolite obtained in pooled urine specimens collected over the same time period (Hammel et al., 2016). In a separate study, she found that the wristbands were equal to or better than spot urine tests at predicting the total mass of several organophosphate ester metabolites excreted in the urine over the course of 5 days (Hoffman et al., 2021).

Silicone Samplers for Companion Animals

To test whether silicone samplers could characterize exposures in pet dogs and to compare the exposures of dogs with those of their owners, Stapleton, Breen, and colleagues studied 30 pairs of owners and dogs over the course of 5 days, with owners wearing silicone wristbands and dogs wearing silicone tags. High-resolution gas chromatography (GC) and mass spectrometry (MS) were used to support the analysis of over 150 target chemicals as well as to support "suspect screening," which allows for identification of nontargeted chemicals that might be present in the specimens. Analysis of both organophosphate esters and pesticides showed significant correlations between dog tag and dog urine, and between wristband and human urine, with tighter correlations observed in dogs (Wise et al., 2020, 2022). The same study found a strong correlation between exposures measured on human wristbands and on companion dog tags for a range of chemicals, including legacy PCBs, a novel flame retardant, pesticides, and a component of flea/tick treatments. Silicone tags have also been successfully used to measure exposures to endocrine-disrupting compounds in cats (Poutasse et al., 2019).

---

[39] Tris(1,3-dichloroisopropyl)phosphate.

Silicone Samplers for Testing Exposure over Time

Stapleton described a recent, unpublished study of 110 pregnant women in New York City, in which the most abundant class of chemicals picked up by targeted screens using these wristbands consisted of phthalates, followed by organophosphate esters. But many additional chemicals were detected using a nontargeted approach. She highlighted a need to support more research on characterizing these mixtures—each individual's "exposure fingerprint"—to support environmental health research and investigate associations between exposure to chemical mixtures and adverse health outcomes, said Stapleton.

Silicone samplers offer many advantages for companion animal research: they are noninvasive, easy to use at home and can be returned via the mail, easy to store, and amenable to both targeted and nontargeted screening approaches, said Stapleton. In addition, they could be used to support prospective research studies and to assess how our chemical exposures change over time. People move around and experience huge seasonal differences in exposures to many chemicals in the indoor environment, and these variations can be captured by silicone samplers. By providing a measure of the integrated average exposure over time, these samplers can help support prospective longitudinal studies. However, it should be noted that they do not capture dietary exposures or individual variation in metabolism of toxicants, said Stapleton. They do not trap metals well, but they might be modifiable to better accumulate particulates that contain metals. Stapleton and Breen have begun to use the tags to study gene–environment interactions associated with an increased risk of bladder cancer in Shelties carrying the BRAF mutation.[40] Citing the work of Kim Anderson's group at Oregon State University, Stapleton noted that silicone tags have the potential to be used in a variety of model organisms, including horses.

Cats as Sentinels for Persistent Organic Pollutants Indoors

Cats have many disease syndromes that closely parallel those in humans, including cat versions of severe acute respiratory syndrome (SARS), acquired immunodeficiency syndrome (AIDS), and other viral diseases, as well as many respiratory and endocrine disorders, said Jan Dye (EPA). As was discussed for dogs, cats' condensed life span, indoor lifestyle, veterinary records, and fully sequenced genome (Pontius et al., 2007) enable their use as sentinels to investigate human disease syndromes in terms of environmental exposures. Dye discussed studies of two persistent organic pollutants in cats: PBDEs and PFAS.

---

[40] See https://www.v.org/grants-awarded/to/matthew-breen-heather-stapleton (accessed on January 3, 2022).

PBDEs and Skinny (Hyperthyroid) Cats

The brominated flame retardants, PBDEs, are endocrine disruptors with structural similarity to thyroid hormones. The first reports of environmental contamination with PBDEs in 1979 coincided with the first reports of feline hyperthyroidism (FH), and increased numbers of FH cases were seen in hot spots for PBDE contamination. PBDEs bioaccumulate in the food chain and are found in fish, which cats eat; they were also used in many household products, including carpets, furniture, and electronics. FH is now the most common endocrine disorder in cats. It is characterized by excess T4 production, which causes weight loss in middle age, and if left untreated will cause heart problems, hypertension, and kidney damage. FH has been proposed as a model for toxic nodular goiter in humans (Peterson, 2014).

The biggest risk factor for FH is living indoors, but certain types of canned cat food are also associated with an increased risk. Dye and colleagues measured PBDE content in cat foods and found a strong correlation between the amount of PBDE that a cat consumed and its likelihood of developing FH (Dye et al., 2007; Martin et al. 2000). The researchers detected multiple different PBDE products accumulating in cat serum at levels 20- to 100-fold higher than median PBDE levels in U.S. adults (Dye et al., 2007; Schecter et al., 2005). PBDE levels were high across all cats and were not significantly different between FH and non-FH cats,[41] though levels correlated with age. Noting that PBDE products tend to leach out of furniture and carpeting, with high levels of PBDEs ending up in house dust that can be ingested by cats during grooming, Dye hypothesized that the very high PBDE levels seen in cats were due in part to this exposure. Like cats, young children inhale, contact, and ingest greater quantities of house dust than adults, due to their increased floor contact time and mouthing behaviors; indeed, the amount of dust ingested by cats falls in the range of that estimated for toddlers[42] and supports the use of cats as sentinels for toddlers, said Dye. Judging from the number of publications that have cited this work, there is widespread interest in this approach, she added.

PFAS and Fat Cats

Given the ubiquitous use of PFAS, which are stain-resistant agents and emerging environmental contaminants, Dye and colleagues performed a cross-sectional analysis of cats to determine whether exposure to PFAS was corre-

---

[41] There was an increase in one specific PBDE, OctaPBDE, in the hyperthyroid cats (Dye et al., 2007).

[42] See https://www.epa.gov/healthresearch/understanding-exposures-childrens-environments (accessed January 4, 2022).

lated with health issues. PFAS are found in many household products, including non-stick cookware, food packaging, furniture, clothes, carpets, paint, and cosmetics. The researchers measured three PFAS compounds (perfluorooctane sulfonic acid [PFOS], perfluorooctanoic acid [PFOA], and perfluorohexane sulfonate [PFHxS]) in house dust, finding them at low levels in most households but very high in some (Strynar and Lindstrom, 2008). Among 70 cats studied, several of the indoor cats (but no feral or outdoor cats) had "amazingly high" serum PFAS levels (Bost et al., 2016), similar to levels previously seen in a case report of human teenagers, whose analyte profile was consistent with the known content of a carpet treatment product (Beesoon et al., 2012). Cats were ranked in quartiles based on total serum PFAS concentrations. Those with the highest levels (quartile 4) were more likely to be overweight and to spend more time indoors (Figure 10). Interestingly, both cats and humans have experienced a similar trend toward increasing obesity since the 1970s,[43] and PFAS is a suspected "obesogen," which describes compounds that interfere with insulin signaling and thyroid signaling and may affect metabolism, though Dye pointed out that the association of PFAS exposure with obesity in indoor cats does not imply causation.

In addition to its association with obesity, PFAS exposure in humans has been linked to kidney and testicular cancer, abnormal thyroid hormone levels, pregnancy-induced hypertension, and high cholesterol with associated liver

---

[43] See https://lanekenworthy.net/2012/05/31/why-the-surge-in-obesity and http://www.catalystcouncil.org/enewsletter/october2011 (both accessed January 4, 2022).

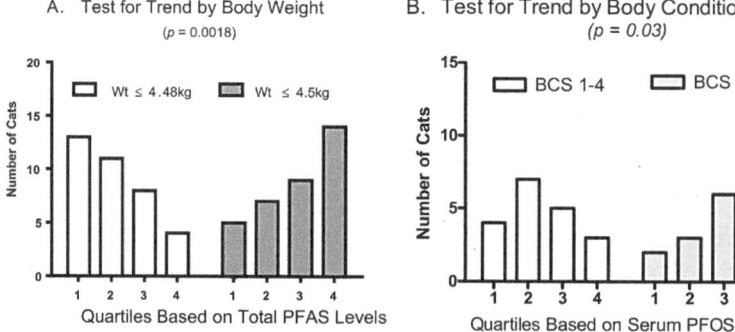

FIGURE 10 Contingency Test for Trend based on (A) associations of body weight and (B) body condition scores with quartile rankings based on serum total PFAS concentrations.
SOURCE: Dye presentation, December 2, 2021. Reprinted with permission from Bost et al., 2016.

disease. When the researchers examined cats with liver disease, feline hyperthyroidism, respiratory effusions, or kidney disease, they found elevated levels of PFAS compounds in these sick cats compared to healthy cats (Bost et al., 2016). "You don't have to look very far into the human literature to realize that obesity, liver, thyroid, and kidney disease are emerging issues that are increasingly being associated with these types of compounds," said Dye, adding that there are many other compounds in the indoor environment. Noting that more than 50 percent of U.S. households have one or more pets, Dye challenged investigators to take collaborative, transdisciplinary approaches to address data gaps in this untapped population and take advantage of the newer technologies, with nontargeted analyses, to identify new associations between exposures and health outcomes.

Radon Exposures and Cancer in Pets

Johannes described research using dogs and cats as sentinels for lung cancer resulting from exposure to household radon. The reported incidence of lung cancer in these pets (where it is called primary pulmonary neoplasia, or PPN) is very low, at 4 per 10,000 dogs in the United States, with an average age at diagnosis of 11 years (Culp and Rebhun, 2020). Secondhand smoke and air pollution are potential environmental risk factors for PPN, and an experimental study in beagles that were exposed to plutonium indicated that radiation could also play a role (Tierney et al., 1996). When Johannes moved from North Carolina to Iowa, which has high levels of household radiation in the form of radon, he noticed an increased number of dogs and cats in his clinics who presented with PPN.

Radon and Lung Cancer in Dogs

Radon is a breakdown product of uranium in the soil that becomes gaseous and circulates in household air.[44] Inhaled radon enters the lung and creates reactive oxidative species that can cause double-stranded breaks in DNA and increase the risk of developing lung cancer (Bersimbaev et al., 2020). As was noted for other indoor contaminants, pet dogs and cats offer an opportunity to study the health effects of radon due to their shorter life span, significant amount of time spent indoors, and absence of compounding occupational exposures. To determine whether a correlation exists between radon and lung tumors in pets, Johannes and colleagues at 10 specialty clinics in universities around the United States calculated the incidence of PPN in dogs and cats living in regions with low, intermediate, and high levels of radon (Fowler et al., 2020). The incidence of PPN in dogs living in high radon zones was more than

---

[44] See https://theoarp.com/What-Is-Radon (accessed January 5, 2022).

twice that of dogs in intermediate or low radon zones, with similar findings in cats. Johannes noted that the reliance on hospital clinics for data skewed the patient population to those able to afford treatment at a specialty hospital, but he added that these same homes would be more likely to have radon mitigation so, if anything, this study might underestimate the true PPN risk. These results suggest related lines of inquiry into the potential role of radon in sinonasal tumors, of which Johannes sees a much higher incidence in pets than in people, as well as its role in dementia and Alzheimer's. Tissue deposition analysis could indicate what regions of the lung or nasal cavity are most exposed to radon.

Johannes noted that veterinary oncology is a relatively new specialty, with the first specialists certified in 1990, and the actual incidence rates for cancers in pets remain uncertain, but he is working with the large diagnostic companies to improve data collection. Getting quality clinical data is another challenge, but the increasing corporate ownership of specialty veterinary practices may lead to greater standardization of medical records and thus provide a better pool of data, he added.[45] A next step would be to perform observational cohort studies that simultaneously study people with their pets. Johannes sees opportunities for public–private research partnerships, for example, by attaching radon monitoring devices to dog collars to signal exposure in the home.

DeGregori noted that, based on extensive studies in humans and mice, radiation exposure seems to primarily lead to selection for cell populations with specific mutations, and this raises the question of how to interpret mutational signatures associated with radon exposure. Birnbaum replied that mutational signatures can be developed for dog tumors, and indeed some are already available. However, to test whether a given chemical is a carcinogen, Birnbaum favors in vitro approaches, key characteristics, and rodent and fish studies; "to test that in the dog is not very helpful," she said.

Heavy Metal Exposures in the Dogs of Chernobyl

The Chernobyl Nuclear Disaster

On April 26, 1986, the Chernobyl nuclear plant in Ukraine experienced an explosion and fires that spewed large quantities of radioactive iodine and cesium into the air and resulted in establishment of a 30-kilometer exclusion zone where no one is permitted to live to this day.[46] Helicopters

---

[45] See https://www.veterinarypracticenews.com/myvpnplus/the-corporatization-of-veterinary-practices-in-america-will-you-be-a-part-of-it (accessed January 5, 2022).

[46] See https://www.who.int/publications/m/item/1986-2016-chernobyl-at-30 (accessed January 5, 2022).

dumped 25 tons of lead powder on and around the reactor, and this lead still contaminates the environment, along with other heavy metals, pesticides, and industrial toxins. Today, thousands of workers visit Chernobyl daily to work on cleanup, remediation, and new construction. Researchers demonstrated a dose-dependent risk of radiation-induced cataracts in the workers who cleaned up Chernobyl in the years immediately following the explosion (Worgul et al., 2007) as well as an elevated frequency of cataracts in the local birds (Mousseau and Møller, 2013), said Norman Kleiman (Columbia University), who observed a dose response for radiation cataracts in Chernobyl's voles.

Toxic Exposures in the Dogs of Chernobyl

Chernobyl is an environmental and ecological disaster as well as a radiological disaster, with multiple industrial waste sites and "contamination everywhere you turn," said Kleiman. Numerous feral dogs populate Chernobyl—descendants of canine survivors after the meltdown—and Kleiman studies the dogs' exposures to toxic and heavy metals, pesticides, and organics, with the hypothesis that the dogs could serve as surrogates for human exposures to the same compounds and provide a measure of the potential risks to human workers. The researchers obtained hair samples from dogs captured throughout the exclusion zone and from a town 40 kilometers outside the zone, as well as from pet dogs elsewhere in Ukraine and in New York City and used inductively coupled plasma mass spectroscopy (ICP-MS) to measure 26 different metals in the samples. They analyzed unwashed and washed hair samples, with the expectation that these measurements would provide information about both external contamination and body burdens of these compounds.

Of the 26 metals tested, 19 were significantly elevated in unwashed hair and 15 in washed hair from dogs living in the exclusion zone compared to unexposed controls, suggesting that exposure was a combination of ingestion/inhalation and external contamination, said Kleiman. Some of these elevations were dramatic—for example, beryllium was increased 66-fold in unwashed hair and 25-fold in washed hair from dogs living near the nuclear power plant compared to those outside the exclusion zone. Kleiman is currently working to correlate the highest exposure levels across the range of metals with a variety of health outcomes. These observations suggest that the heavy metals in the environment around the power plant could pose health hazards to both workers and animal populations and support the use of dog hair samples and other biomarkers as surrogate indicators for human risk at the power plant and other sites of industrial and radiological disasters, said Kleiman.

Exposures through Food

Dogs as Models for Human Dietary Exposures

Given the extent to which dogs recapitulate human exposures, they constitute the perfect model for testing nutritional interventions for humans, said Joseph Wakshlag (Cornell University), although he noted that the diet and eating schedules of humans differ greatly from those of companion dogs. Thanks to federal regulations, commercial dog and cat foods offer a complete spectrum of vitamins and minerals, but they also tend to be higher in fat and protein than the human diet, with more long-chain omega-3 fatty acids. Nonetheless, the diets of dogs, despite being contained in a dry pellet, resemble those of humans better than the casein-and-lard-based rodent chow fed to lab mice and rats. Birnbaum noted that the outcomes of nutrition studies in rodents were affected by the sources of the proteins, carbohydrates, and other nutrients, and Wakshlag said that the longitudinal dog studies will address some of these questions. He said that for canine cancer studies, it is very important to control the diet, noting that all dogs enrolled in the Vaika Aging Project,[47] which tests an antiaging intervention in retired sled dogs, receive the same complete and balanced kibble. Finch noted that the sources of protein in dog food are commodities driven and can vary widely with the season and international markets. The source of fiber is also significant, said Elizabeth Ryan (Colorado State University); beet pulp and rice bran can both contribute to the percent fiber on a label, but they contain different prebiotics and will impact the microbiome differently. Fiber also affects how much protein becomes bioavailable, while fat influences the bioavailability of lipid-soluble vitamins. It will be important to study nutrient availability in these diets, added Ryan.

Heterocyclic Amines and Acrylamide in Food

Heterocyclic amines, which form adducts that are mutagenic and carcinogenic (Wakabayashi et al., 1992), build up in the hair of human omnivores (but not vegetarians) and to varying degrees in dogs (Gu et al., 2012). The intake of heterocyclic amines by dogs was estimated to exceed that of humans by a factor of five, and when these compounds were extracted from commercial pet foods, they were found to have mutagenic activity in laboratory assays; of 25 pet foods tested, 24 were mutagenic (Knize et al., 2003). Commercial dog food provides a complete and balanced diet that avoids nutrient deficiencies, but dried food production is also likely to generate heterocyclic amines because pet foods are manufactured through a process of heating, extrusion,

---

[47] See https://www.vaika.org/the-project (accessed January 6, 2022).

and dehydration to produce a food that is dry and stable, said Wakshlag. Vegetarian dog foods are becoming increasingly popular, noted Wakshlag, with protein levels similar to meat-based foods, so comparing dogs on the two types of food could shed light on the effects of processing.

Commercial dog and cat foods also contain the neurotoxin acrylamide, as do fried potatoes that are commonly consumed by humans.[48] Acrylamide can be metabolized into the highly potent mutagen glycidamine and is listed as a carcinogen in the state of California, which sets the no-significant-risk level (NSRL) at 0.2 μg/day.[49] When Wakschlag and colleagues tested dog foods for the presence of acrylamide, which is more concentrated in dry than wet foods, the dry food with the highest level provided a dose of roughly 34 μg/day, far in excess of the NSRL but comparable to the exposure of a person eating French fries daily. Wakshlag said it will be interesting to compare diet-related cancer risks as fresher, minimally processed dog foods enter the market.

Heavy Metals in Food

The levels of heavy metals in pet food are thought to be relatively safe, said Wakshlag, though dry dog foods approach Conservative Toxicity Reference values for cadmium and arsenic (Macías-Montes et al., 2021). Levels are dependent on the protein source, with higher levels of mercury and arsenic in fish-based kibble than in meat-based or poultry-based kibble (Kim et al., 2018). He suggested developing a repository of data concerning the composition of commercial dog foods that includes not just nutrients but also heavy metals and heterocyclic amines.

Chemical Mixtures, Cancer, and Diet—The Example of Pesticides

Cancers are characterized by cellular hallmarks that reflect the transitions from tumor initiation to tissue invasion and metastasis (Goodson et al., 2015), said Elizabeth Ryan (Colorado State University). Noncarcinogenic pollutants in the environment can affect these hallmarks at low doses, leading to the hypothesis that the cumulative effects of combined exposures to multiple agents at low doses may promote tumor progression (see Figure 11). Pesticides exert harmful effects through oxidative stress, which causes DNA damage, and by modulating gene expression, possibly through epigenetic alterations, and thus pose many dangers to humans, including increasing the risk of cancer,

---

[48] See https://www.researchgate.net/publication/355392171_Beynen_AC_2021_Acrylamide_in_petfood (accessed March 29, 2022).

[49] See https://oehha.ca.gov/proposition-65/proposition-65-list (accessed January 5, 2022).

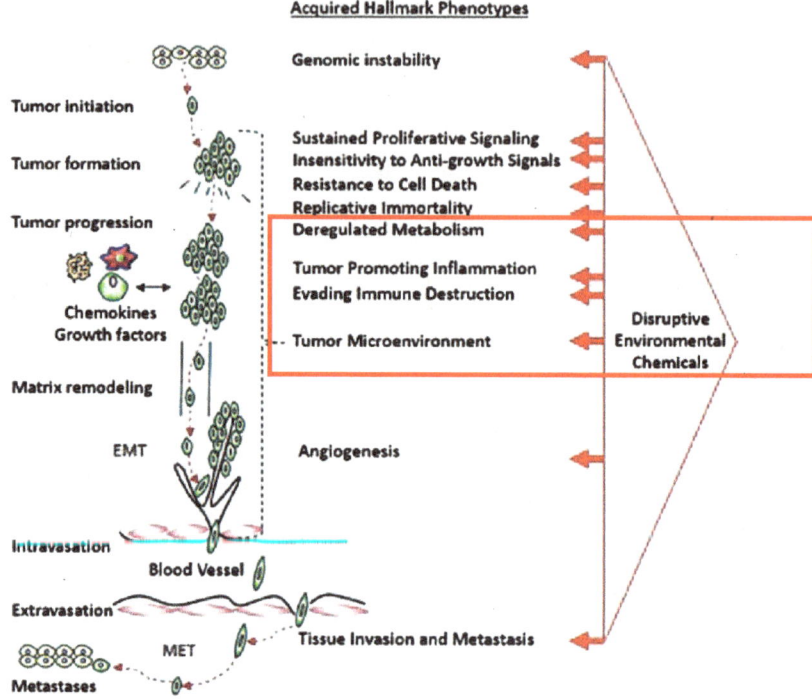

FIGURE 11 The Hallmark Framework of cancer progression.
Many pesticides are known to perturb pathways in this progression, with well-characterized mechanisms and endpoints. Four hallmarks, shown in the box, offer particular opportunities for use of the dog model to advance knowledge about pesticides.
NOTE: EMT = epithelial mesenchymal transition; MET = mesenchymal to epithelial transition.
SOURCE: Ryan presentation, December 2, 2021. Adapted from Goodson et al., 2015.

she said. Pesticides are often present in combination with other toxicants in the environment, where exposure occurs through contact with contaminated drinking water or food, said Ryan. Dogs drink large quantities of water both in the home and outside, exposing them to these mixtures (Aslan et al., 2020). Stapleton noted that although spot urine is variable for many chemicals, a lot of pesticides, phenolic compounds, and amines could be studied in dogs' spot urine samples, in connection to their potential risks for adverse health outcomes later in life.

Metabolism May Modulate Exposure Risk

When studying the effects of complex, low-dose chemical exposures such as those from pesticides, it is important to take into account metabolic differences among animals, which may influence exposure dose, immunity, and risks for developing cancer, said Ryan. For example, there are metabolic differences between small and large dogs. Among laboratory dogs that were housed together and fed identical diets, small dog breeds could be distinguished from large breeds based on distinct plasma metabolome signatures and distinct clinical signatures; there were also differences in the fecal microbiome (Middleton et al., 2017). Body condition also has an impact on metabolism, and potentially on exposure: When Ryan and colleagues fed an identical diet to 66 companion dogs that were healthy weight, overweight, or obese, the dogs' plasma, fecal, and urine metabolomes differed based on body condition score (Forster et al., 2018). Plasma lipid composition was altered in the overweight and obese dogs, indicating that, as in humans, being obese/overweight affects metabolism in dogs and may impact levels of pesticides, said Ryan. Indeed, in a 1989 study on companion dogs that found an association between insecticide exposure and bladder cancer risk, this risk was enhanced in overweight or obese dogs (Glickman et al., 1989).

Diet May Modulate Exposure Risk

Diet also has the ability to modulate exposure, and this may derive in part from its effect on the microbiome. When Ryan and colleagues screened 21 healthy-weight dogs for 301 pesticides in their urine, they detected atrazine in all 21 dogs and combinations of multiple pesticides in 20 (Forster, et al., 2014). Changing to a 25 percent bean diet altered the dogs' metabolomes (Forster et al., 2012, 2015) as well as their fecal microbiomes (Kerr et al., 2013). Ryan noted that chemicals from pesticides interact bidirectionally with the gut microbiota; on the one hand, microbial enzymes can alter these chemicals, potentially leading to activation or detoxification, and on the other hand, pesticides can contribute to dysbiosis (reduction in microbial diversity) (Giambò et al., 2021). Antibiotics can increase pesticide bioavailability (Zhan et al., 2018), indicating that disruptions to the gut microbiome may increase the risk from pesticides and highlighting the need to study the effect of medications on pesticide risk, said Ryan.

As a proof-of-concept for testing the ability of the microbiome to mediate the disease-causing effects of environmental chemicals, Ryan and colleagues examined the effect of diet change on the microbiota of dogs undergoing chemotherapy for cancer. Lymphoma alters the fecal microbiome of dogs (Gavazza et al., 2018), as does chemotherapy, she said. When dogs undergoing chemotherapy were fed a diet of 20 percent beans and 5 percent rice bran, some microbial diversity was restored to the gut and probiotic commensals increased, suggesting that it may be possible to modulate the effects of environmental exposures through dietary modifications, said Ryan. However, it remains unclear what constitutes a healthy microbiome in dogs, or for that matter in humans, which makes it difficult to assess the impacts of dysbiosis. Nonetheless, there are similarities in human and dog gut microbiomes, and microbes are exchanged between humans, animals, and the environment (Trinh et al., 2018). Metabolomics offers a powerful tool for understanding the global landscape of human gut microbiota (Sung et al., 2017) and can be used for this purpose in dogs, added Ryan.

Chemicals in Flea-and-Tick-Control Products

The mechanisms of action of many pesticides used for flea and tick control are connected to pathways involved in cancer, said Ryan. Participants considered the relative benefits of controlling fleas and ticks in dogs, versus the risks of exposing dogs and humans to pesticides. "Dogs are constantly being bombarded with these pesticides" that carry a potential risk of cancer, said Wakshlag, to which Johannes replied that, from an infectious disease standpoint, keeping ticks and fleas off dogs has a tremendous quality-of-life benefit, and it may come down to weighing the trade-offs. Nontoxic alternatives were mentioned, but evidence of their efficacy is lacking.

The activities of topical flea and tick treatments could be altered by the skin microbiome, said Ryan, and participants considered studying the skin microbiome as a reflection of exposures. While dogs' fur makes collection of skin microbes a challenge, Ryan remarked that techniques for collection of the skin microbiome have evolved, and it would be valuable to develop a standardized sampling scheme. Some animals also develop allergies to these chemicals, and age-related susceptibility to allergy may be an untapped indicator of environmental compounds acting on the immune system.

Discussion: Relevance of Companion Animal Exposures to Humans

Communicating Exposure Risks to the Public

Workshop participants considered the need and complexity of communicating exposure information to the public, particularly when researchers do not yet understand the relationship between a given exposure and disease because so much of the data from companion animals are correlative. The exposure–disease connection needs to be something the general public can understand, said Miller. Scientists are interested in molecular mechanisms but what really matters are the functional consequences, and these have to be actionable. Reducing plasticizers or changing diet are actionable goals, "and if it doesn't get to that translation point, we're just wasting our time," he added. This requires help from science communicators. For example, DNA testing companies put considerable thought into what to tell a person with a particular polymorphism.[50] "We really have to think about that message part at the end," said Miller. Deziel added that communication with policy makers is also needed to drive action on public health. Stapleton said she reports measured exposures back to her study participants, and she stressed the importance of providing context. Her reporting template indicates whether a particular exposure is above or below the median, provides guidance for reducing exposure, and makes it clear that the researchers still do not know the risk associated with the exposure.

Deconstructing Complicated Environmental Exposures—Or Not

We live in a "chemical soup" together with our dogs, said Birnbaum. Rather than trying to deconstruct the mixtures, maybe we could look on a larger scale—diet, pesticides, air pollution—and not fixate on knowing everything in the mixture. Exhaustive efforts to identify carcinogens in cigarette smoke identified a group of "proven" and "possible" carcinogens, noted Finch. Yet, remarkably, cigarette smoke and air pollution do almost the same thing despite the huge range of variation, so singling out individual components would not be terribly useful. Approaching the problem on a macro level—for example, analyzing how diet and omega-3 content may protect against Alzheimer's disease in relationship to environmental factors that are not yet defined—may prove the most rewarding, said Finch.

Dogs as Models for Interventional Studies

Dogs are a remarkable, and yet untapped, resource for interventional studies of cancer, said Wakshlag. Their nutrient intake is far easier to control than that of humans and ripe for studying the influence of the omnivore diet on both cancer prevention and progression. However, prevention studies are currently hamstrung by the fact that little is known about the composition of commercial dog food other than macronutrient content. To get the necessary information on heavy-metal exposure, pesticides, and so forth, when there are 250–300 companies, each with 5–40 brands on the market, is a "gargantuan

---

[50] See https://www.23andme.com (accessed January 13, 2022).

task," but no manufacturer will fund this work, so it needs government support, said Wakshlag.

Johannes emphasized the need to partner with industry. He noted that state-of-the-art diagnostics are now available to veterinarians, making it possible to match environmental exposures to mutations and biomarkers. However, the therapeutics have lagged behind, so even when drugs exist for human cancers, veterinarians are often unable to use them because they have not been validated in dogs or cats. There is an opportunity to test mitigation strategies via parallel studies in dogs, said Page. This might include, for example, dietary measures to mitigate exposures to heavy metals or air pollution.

Prioritizing Exposures for Future Research

Given the broad range of exposures that could be measured and finite resources, prioritizing the questions to study will be crucial, said Promislow, who noted that studies vary in their aims. There are banked collections around the world and new collections being started (see section on Biobanks for Companion Animal Sentinel Studies). Some questions address widespread conditions and could potentially benefit many, whereas others address rarer conditions. How can the research be optimized? Stapleton envisioned focusing on ambient measurements that are cumulative over time, in an exposomic framework, while at the same time paying careful attention to the diet and microbiome. Ryan highlighted changes to the microbiota and immune evasion by tumors as two potential readouts of the effects of chemical mixtures.

The key value of these longitudinal studies is their tissue and data banks, said Wendy Shelton (Virtual Beast/Colorado State University), and it will be important for future researchers with fresh ideas and technologies to be able to access these resources. Ten years from now, when the results come in from prospective dog cohorts, we'll be able to go back and look at specimens to determine the relationships between priority agents and disease, said Ryan. The question will then be whether to respond by designing an intervention, she added.

Wakshlag said that given the capability to measure "just about anything" in interstitial fluid and the ongoing longitudinal dog cohorts, now is the time to assemble expert panels to identify the important exposures to study, and to evaluate the risks of pesticide exposures based on available human and dog data. Considering his own data on heterocyclic amine and acrylamide exposure in dogs and humans and their potential long-term health risks, it is also time to reconsider the established safe upper limits for these compounds, he added. Study design, statistical analysis, and readout also require consideration, said Ryan.

ACCELERATING CROSS-SPECIES COMPARISONS: OPPORTUNITIES AND CHALLENGES IN DATA SOURCES, COLLECTION, STORAGE, MODELING, AND SHARING

Participants discussed practical concerns for collecting and storing data and making it available to the wider research community, as well as methods for comparing data obtained from companion animals to that collected from humans. Speaker presentations considered strategies for standardizing, sharing, and aggregating health records and relevant metadata across species; the obstacles (e.g., scientific, infrastructure, ethical, and financial) to using companion animals as biomonitors; incentives needed to encourage data sharing and collaboration; and implications for expanded, systematic collection of these data.

Human Exposure Assessment

Widespread ambient environmental exposures, ranging from air pollution and drinking water contamination to nonchemical agents such as light pollution, may affect health even at low levels, said Rena Jones (NCI), who is building data resources to support cancer epidemiology studies of these exposures in humans. Jones combines individual and population exposure data, collected through environmental monitoring, biological sampling, and surveys, with secondary data derived from exposure assessments that were not specifically intended for epidemiological studies, such as regulatory monitoring measurements, census data, and satellite imagery; these data are then linked to individuals based on geographic location, using geographic information systems (GIS).

Use of Spatially Derived Exposure Assessments to Predict Individual Exposures

Jones described two studies that used spatially derived exposure assessments to predict individual exposures. In one study, Jones and colleagues probed the association between lung cancer and ultrafine particles (UFPs)—components of outdoors air pollution that are derived from combustion of diesel fuel and behave as a gas, passing through walls and cell membranes. For the Los Angeles Ultrafines Study, Jones and colleagues obtained health data from a cohort of 52,000 participants enrolled in an existing longitudinal study.[51] To determine UFP exposure, the researchers measured UFPs and other pollutants at 250 monitoring sites in the Los Angeles basin, detecting a range of exposures that were then analyzed by linear modeling called land use regression, which considered nearby traffic, proximity to airports, population density, and other location-based environmental characteristics. This generated an equation that could be used to predict UFPs at each participant's location, and the predictions were validated against measurements taken at a subset of cohort residences (Jones et al., 2020).

In a second study, which assessed exposures to nitrate in residential well water, Jones and colleagues applied a different GIS approach—random forest modeling—to analyze nitrate measurements that had been obtained previously through community and personal well-water sampling in rural Iowa and North Carolina.[52] Nitrate is an agricultural contaminant that is regulated in public drinking water but not in private wells. The researchers identified 58 predictive variables, which were validated against sampling with good accuracy and high specificity (Messier et al., 2019), enabling them to predict nitrate levels in private wells at levels below the regulatory limit.

Use of Regulatory Data to Predict Exposures

In ongoing work to test the feasibility of estimating exposure to PFAS from drinking water, Jones and colleagues are applying GIS analysis to a combination of regulatory data from the EPA that measures PFAS levels in the public water utilities[53] and participants' responses to detailed surveys about their water intake, to identify determinants of serum PFAS in 1,200 women nested within a cohort in California.[54] Jones and colleagues also use regulatory data to estimate exposures to the nearly 800 chemicals, released from over 15,000 industrial point sources scattered around the country, that are reported to the EPA each year under the Toxics Release Inventory (TRI) program.[55] Half of these point sources emit known or probable carcinogens. To turn a TRI emissions estimate into an exposure estimate, the researchers employ inverse distance weighting, summing the emissions from all facilities within a given radius of a participant and weighting each one by the inverse of its distance to the participant, with additional fine-tuning (e.g., adjustment for prevailing wind direction). Jones noted that these estimates

---

[51] See https://dietandhealth.cancer.gov/history.html (accessed January 18, 2022).

[52] See https://aghealth.nih.gov (accessed January 18, 2022).

[53] See https://www.epa.gov/dwucmr (accessed January 18, 2022).

[54] See https://www.calteachersstudy.org (accessed January 18, 2022).

[55] See https://www.epa.gov/toxics-release-inventory-tri-program (accessed January 18, 2022).

are not surrogates for exposure, but they enable ranking of participants based on their relative exposures. Focusing on dioxin and PCB levels in serum obtained from the NHANES, the researchers validated these GIS-derived exposure estimates against actual burden as measured in serum. Adding satellite imagery allows for triangulation of location data, to (for example) determine the height of a smokestack, enabling the construction of actual pollution dispersion models.

Jones noted that these same types of exposure assessment approaches centered on residential locations can be applied to the study of companion animals as sentinels for humans, with the advantage that the shorter latencies and life spans of animals require a shorter monitoring period. Furthermore, although focusing on exposures in the home may be an oversimplification for humans, who move around, it may be sufficient for pets. Jones encouraged companion animal researchers to propose ancillary studies to human studies that are recruiting new cohorts, such as the intramural NCI Connect for Cancer Prevention study,[56] which includes infrastructure for GIS and location-based exposure assessments, with complete residence histories and drinking water intake data. She expressed the need to gather more data on disadvantaged populations, which tend to be underrepresented in many cancer cohorts.

Comparative Oncology: How Dogs are Helping Researchers Understand and Treat Cancer

Unlike laboratory models of disease, natural development of cancer in companion dogs offers "the really unique advantage of having a coevolving tumor, stroma, immune system, and microenvironment," and dogs can be used to evaluate novel therapies that cannot be tested in treatment-naïve human disease, said LeBlanc (LeBlanc and Mazcko, 2020). Following a 2015 workshop on comparative oncology (NASEM, 2015), there has been a groundswell of research in this field, aided by a network of veterinary schools that conduct clinical trials through the NCI Comparative Oncology Trials Consortium (COTC).[57] This has driven a dramatic improvement in data to inform drug development and advance the translational relevance of canine cancer for humans, said LeBlanc, whose Comparative Oncology Program engages in laboratory research and COTC-run clinical trials and supports additional

---

[56] See https://www.cancer.gov/connect-prevention-study (accessed March 7, 2022).

[57] See https://ccr.cancer.gov/comparative-oncology-program/consortium (accessed January 18, 2022).

collaborations, such as the ICDC and a positron emission tomography–computed tomography (PET/CT) and Molecular Imaging Consortium.[58]

The Companion Dog Model for Osteosarcoma

Osteosarcoma is diagnosed in about a thousand children or young adults per year in the United States, and this rarity makes it a challenge to study in humans. The 5-year survival rate for children with metastatic osteosarcoma is less than 30 percent, and treatments have not improved in 30 years, despite significant efforts. New drug targets are needed, said LeBlanc (Bishop et al., 2016). Companion dogs are good models for studying osteosarcoma because it is diagnosed in more than 10,000 dogs each year, she added. Most canine osteosarcomas are metastatic, with a survival time of about 1 year (Selmic et al., 2014). Pediatric and canine osteosarcomas look strikingly similar at every level: histopathological, molecular, and clinical, including the tendency of the tumor to metastasize to the lung (Gustafson et al., 2018). There are broader similarities as well, with osteosarcoma affecting big dogs and, in general, tall children. There may be a baseline risk that is modified through the onset of rapid scalable growth through puberty, with the effects of estrogen on bone exerting a protective effect on girls, who are affected less frequently than boys (Simpson et al., 2017), though this is harder to assess in dogs that are spayed or neutered. The clinical similarities in the human and dog diseases support hypotheses involving IGF pathways, micro stresses, micro fracture/trauma, or bone remodeling, said LeBlanc.

LeBlanc and colleagues are working to define a genomic landscape for canine osteosarcoma that can be integrated with human osteosarcoma data to identify new druggable targets for preclinical work in mice and then dogs, with the ultimate goal of benefitting both canine and human patients. Unlike humans who enroll in clinical trials of novel therapies, dogs with osteosarcoma provide a source of treatment-naïve tissue, enabling LeBlanc and colleagues to develop a biospecimen repository with primary tumors, matched normal tissue, blood, and other samples. The researchers are using these specimens to describe the molecular landscape of canine osteosarcoma, in a project called Deciphering the Osteosarcoma Genome in Dogs (dog^2) (LeBlanc and Mazcko, 2020) (see Figure 12).

Using Dog Osteosarcoma Data to Inform Human Medicine

With the goal of using canine molecular and clinical data to inform development of therapies for osteosarcoma, researchers conducted clinical trials for

---

[58] See https://ccr.cancer.gov/comparative-oncology-program (accessed January 18, 2022).

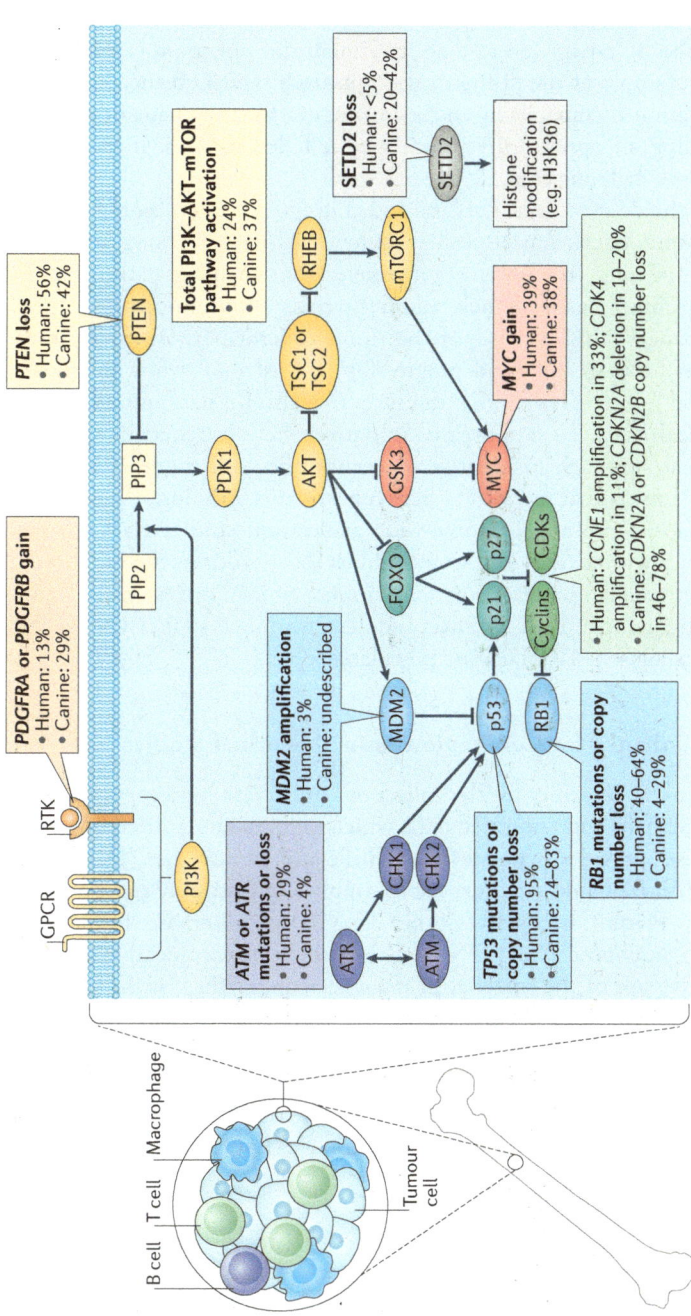

FIGURE 12 Comparative molecular features of canine and human osteosarcomas.
There are many similarities in molecular pathways and genetic changes that occur in dogs and humans; osteosarcomas are shown.
SOURCES: LeBlanc presentation, December 2, 2021. From LeBlanc and Mazcko, 2020.

dogs testing two different drug treatments and obtaining biospecimens from about 450 canine osteosarcoma patients. Their first published study, which found that adding rapamycin to dogs' treatment did not improve outcomes, provides an example of the utility of this approach, said LeBlanc, informing the consideration of rapamycin and its analogues for treatment of children while providing an opportunity to ask why it failed to work in dogs on a molecular level (LeBlanc et al., 2021).

Within the dog^2 project, LeBlanc and colleagues are combining analysis of bulk genomic data from clinical specimens; AI/deep learning and computational modeling to identify biomarkers from imaging data; immune landscape profiling, using a new canine-specific immuno-oncology gene expression panel;[59] and analysis of the tumor microenvironment to interrogate tumor heterogeneity and progression, in order to define prognostic signatures and tumor subtypes that can form the basis for harmonized canine/human clinical trials. To support this initiative, the researchers developed a computational approach to identify cancer-specific gene modules and biological mechanisms that are shared between humans and dogs (Tawa et al., 2021). Perhaps combining genome-wide association studies (GWAS) and tumor genetics with activity tracking could elegantly address the question of whether osteosarcoma requires a certain number of load cycles on the limbs. If so, this might explain why it affects adolescent humans (ages 12–16) and geriatric dogs (ages 8–10), LeBlanc postulated.

Biobanks for Companion Animal Sentinel Studies

A biobank is "a facility for the collection, preservation, storage, and supply of biospecimens and associated data, which follows standardized operating procedures and provides material for scientific and clinical use" (Hewitt and Watson, 2013). Biobanks are intended to supply and distribute specimens for research, said Marta Castelhano (Cornell University), otherwise they become "biohoards" (Catchpoole, 2016). Nonetheless, there are considerable obstacles to the effective use of the material in many biobanks. These include lack of uniform systems for banking, data annotation, and bioinformatics, and insufficient or variable consent procedures (e.g., limiting analysis to a specific set of environmental agents). There is also often reluctance to share biospecimens and difficulty in accessing them from outside of the institution, and many existing biobanks lack the necessary information to link participants to exposure data.

---

[59] See https://www.nanostring.com/products/ncounter-assays-panels/oncology/canine-io (accessed January 18, 2022).

Standards for Biobanks

The first international standard for biobanks, ISO 20387, was published in 2018,[60] and two documents offer best practices to guide the collection, storage, and analysis of biobank specimens (Campbell et al., 2018; NCI, 2016). Castelhano emphasized the importance of conformity assessment (CA), which demonstrates that a biobank complies with specific requirements. CA can be performed at three levels of stringency: first party, where the biobank attests to its own conformity; second party, where a user reports on conformity; or third party, where an impartial external body provides certification or accreditation, with accreditation going the farthest by evaluating the technical competence of personnel (Mouttham et al., 2021). Biobanks should aim for the most stringent assessment practice with accreditation, said Castelhano, saying that ten billion dollars a year are lost to irreproducible research in the United States due to inadequate biological reagents and reference materials (Freedman et al., 2015). She emphasized that "standardization and auditing of biological materials . . . can enhance cumulative production of scientific knowledge by improving both availability and reliability of research inputs" (Freedman et al., 2015; Furman and Stern, 2011). In addition to assuring technical competence, said Castelhano, accreditation enables harmonization of procedures, allowing for interoperability among biobanks. This is especially important when multiple institutions are involved, as is often the case for exposure studies (Lermen et al., 2020).

Biobanks for Companion Animal Sentinel Studies in the United States

Established U.S. biobanks offer opportunities to advance research on the role of companion animals as sentinels for environmental exposure, said Castelhano. One of these is the National Institutes of Standards and Technology (NIST) biorepository in Charleston, South Carolina, which contains two environmental specimen collections: the National Biomonitoring Specimen Bank, established in 1979 to store human specimens, and the Marine Environmental Specimen Bank, established in 2002 to store diverse marine specimens (Becker and Wise, 2006). Both biobanks were designed for long-term storage and use over the next 50–100 years. The Centers for Research in Emerging Infectious Diseases (CREID)[61] is creating an international, standardized biobanking infrastructure in several countries, specifically for pandemic pre-

---

[60] See https://www.iso.org/obp/ui/#iso:std:iso:20387:ed-1:v1:en (accessed January 18, 2022).

[61] See https://creid-network.org (accessed January 18, 2022).

vention, which includes the collection of companion animal specimens, said Castelhano.

Castelhano coordinated the biobanking of high-quality, clinically annotated specimens from the Dog Aging Project, with the intent of making them available to the wider research community. As noted in the previous discussion of the Dog Aging Project, specimens are collected from thousands of dogs by veterinarians across the United States, sent to Texas A&M, and then shipped to Cornell, where they are catalogued in the Biobank Information Management System, processed, and stored. Whole blood specimens are processed into serum, plasma, and live peripheral blood mononuclear cells (PBMCs). Castelhano noted that while blood and urine are easy to obtain, they reflect recent exposure rather than long-term accumulation. Due to the relatively low levels of environmental contaminants in these fluids, large volumes are required to provide sufficient material for biobanking. In contrast, postmortem tissue offers greater potential for assessing bioaccumulation. Postmortem specimens from the Cornell Veterinary Biobank, collected by its rapid necropsy team with an average postmortem interval (time from death to specimen preservation) of 43 minutes, include multiple tissues "from brain to bone" as well as histopathology authentication. These specimens, along with clinical data, are available for use by the larger research community, said Castelhano.

To encourage specimen sharing and collaboration, biobanks need infrastructure support, including HEPA-filtered air in sample-preparation and freezer rooms; cryogenic homogenization; and 24-hour monitoring systems for freezers; as well as risk management plans to avoid loss of collections during a natural disaster, said Castelhano. Standardized protocols are crucial to ensure specimen integrity and stability during a long storage, and biospecimens should be routinely analyzed to assess stability. Standardized consent procedures are also essential. Researchers need a clear access policy, procedures for locating specimens and data, and standardized annotations, including geo-exposure data. Ideally, Castelhano said, she would like to see a national, centralized environmental specimen banking initiative for human and sentinel species. She suggested that researchers establishing biobanks engage three organizations with specimen and data locator capabilities: the Clinical and Translational Science Award One Health Alliance (COHA),[62] the International Society for Biological and Environmental Repositories (ISBER),[63] and the NIST.[64] She also suggested using the model established by BBMRI-ERIC,[65]

---

[62] See https://www.ctsaonehealthalliance.org (accessed January 18, 2022).

[63] See https://www.isber.org/page/about (accessed January 18, 2022).

[64] See https://www.nist.gov (accessed January 18, 2022).

[65] BBMRI-ERIC is a European research infrastructure for biobanking. See https://www.bbmri-eric.eu (accessed January 18, 2022).

a distributed research infrastructure for biobanks in Europe that includes compliance audits, and added that the Cornell Veterinary Biobank accepts materials from smaller studies.

Using Ontologies to Unify Genomics and Phenomics across Species

The human genome has been sequenced, but researchers still do not know much about what it does, how the environment affects gene expression, or how these phenotypes manifest themselves, said Thessen. Adding more species to a data set improves the information landscape. For example, although only about 4,000 of the 19,000 known human genes have gene-to-phenotype annotations, assigning the annotations from five model organism species to their human orthologs boosts phenotypic coverage to 16,000 human genes. Noting that every species has unique phenotypes that optimize it for a different research purpose, from snails modeling spontaneous regeneration of injured neurons to armadillos acting as the only nonhuman host of *M. leprae*, Thessen draws on a wide range of organisms to construct two types of data infrastructure: ontologies and knowledge graphs.

Developing Ontologies across Species

An ontology is a formal, computational representation of knowledge in a particular domain, which uses defined terms with clearly defined hierarchical relationships (e.g., an efferent neuron is a type of neuron, which is a type of neural cell, which is a type of cell).[66] Each relationship between terms is accompanied by a series of formal logic rules readable by software "reasoners," enabling logical inferences and sophisticated data queries. The challenge in developing a multispecies ontology is that each species' genomic database employs a unique vocabulary to describe phenotypes (Thessen et al., 2020). For example, "ulcerated paws" in the mouse corresponds to "palmoplantar hyperkeratosis" in humans and "hairy dog feet" in dogs, so software could not connect the three. Instead, Thessen and colleagues break down complex phenotypes into basic, species-neutral elements, producing, for example, "increased keratinization of stratum corneum layer of the skin on the autopod," which could describe this particular phenotype across all three species (Figure 13). The researchers then use a unifying ontology to combine multiple species in the same ontology and thereby generate testable hypotheses. For example, in the unifying phenotype ontology, UPheno,[67] "enlarged heart

---

[66] See bit.ly/ontology101 (accessed January 19, 2022).
[67] See https://obofoundry.org/ontology/upheno.html (accessed January 19, 2022).

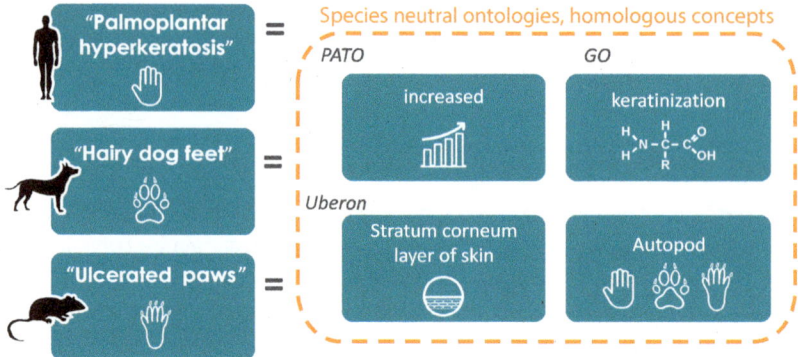

FIGURE 13 Development of species-agnostic ontologies to classify phenotypes across species.
In this example, terms from the Phenotype and Trait Ontology (PATO),[a] the Gene Ontology (GO),[b] and an anatomy ontology (Uberon)[c] are combined using relationships that adhere to formal logic rules to describe equivalent phenotypes in humans, dogs, and mice, enabling these phenotypes to be grouped by computational methods into a unifying ontology.

[a] See https://www.ebi.ac.uk/ols/ontologies/pato (accessed February 11, 2022).
[b] See http://geneontology.org (accessed February 11, 2022).
[c] See https://uberon.github.io/about.html (accessed February 11, 2022).
SOURCE: Thessen presentation, December 2, 2021.

in zebrafish" and "enlarged heart in human" are both subtypes of "enlarged heart," so the human orthologues of genes that impact heart size in zebrafish might be hypothesized to influence heart size in humans, said Thessen.

Knowledge Graphs Map Data across Ontologies

As the data infrastructure grows more complex, it enters the territory of a knowledge graph, such as that of the Monarch Initiative (Shefchek et al., 2020).[68] The knowledge graph integrates genotype–phenotype data across species, mapping data from multiple sources to a small number of unifying ontologies (e.g., UPheno) and combining these unifying ontologies. The Monarch knowledge graph has helped diagnose rare diseases that manifest differently in each patient, using fuzzy matching over groups of phenotypes. It can also help prioritize candidate genes for disease diagnosis or treatment (Smedley et al., 2016).

The missing piece of the knowledge graph is exposures, said Thessen, who is working with exposure scientists and toxicologists to create a semantic data

[68] See https://monarchinitiative.org (accessed January 19, 2022).

model for representing exposure data. Each exposure can be modeled as an event that joins a stressor/stimulus and a receptor (individual, part, or group of organisms); the exposure has an outcome (phenotype or disease); and each of these elements—event, stressor, receptor, and outcome—contains its own set of metadata (Mattingly et al., 2012). The exposure ontology can then be integrated into the larger knowledge graph, connecting exposures and outcomes to genes and phenotypes across species. Thessen gave an example of a zebrafish exposure to a toxin, aldicarb, which promotes microcephaly, which is affected by the rpl11 gene (Figure 14).[69] The human rpl11 ortholog is connected to multiple phenotypes and diseases, raising the hypothesis that rpl11 may mediate disease caused by aldicarb exposure in humans. Dogs have a rpl11 ortholog, though standardized dog and cat phenotype and disease data sets are lacking; "we need a dog phenotype ontology," said Thessen. The knowledge graph will offer the exciting opportunity to do "fuzzy matching" using exposures, genes, and phenotypes, bringing together heterogeneous data from multiple species and modalities. Thessen noted that while chemical exposures are straightforward to describe in machine-readable language, some exposures, including many related to socioeconomic status, are hard to define semantically, and individuals generally experience multiple exposures simultaneously.

Data, Samples, and Modeling

The Mars Petcare Science Engine

Mars Petcare has anonymized data on nearly a hundred million unique pets through its global network of brands in over 125 countries including a number of pet food brands,[70] more than 2,500 veterinary hospitals and diagnostic labs, DNA testing services,[71] and pet activity trackers,[72] said Angela Hughes (Mars Petcare). To enable robust analysis of the large amount of collected data, which includes granular information on genetics, health, nutrition, and behavior, researchers at Mars Petcare's Waltham Petcare Science Institute built the Petcare Science Engine, a cloud-based platform.[73] The Petcare Science Engine inputs data from electronic medical records in both structured and unstructured formats and uses artificial intelligence to structure the unstructured text (medical notes) for analysis. Mars Petcare's massive

---

[69] See https://ntp.niehs.nih.gov/whatwestudy/niceatm/test-method-evaluations/dev-tox/seazit/index.html (accessed February 11, 2022).

[70] See https://www.mars.com/made-by-mars/petcare (accessed March 8, 2022).

[71] See https://www.wisdompanel.com/en-us (accessed January 20, 2022).

[72] See https://www.whistle.com (accessed January 20, 2022).

[73] See https://www.youtube.com/watch?v=UanTIsbMY_8 (accessed January 20, 2022).

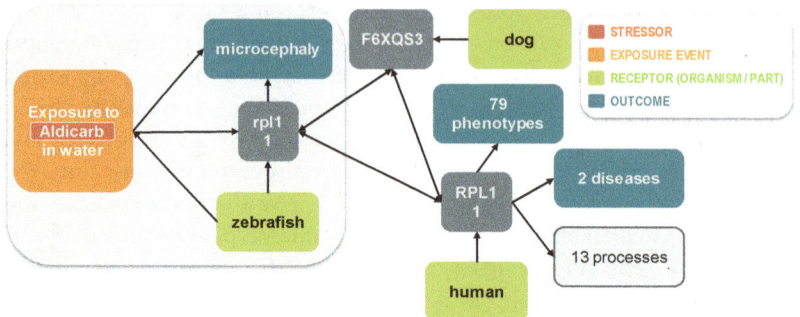

FIGURE 14 Integration of exposure event modeling with the Monarch Knowledge Graph.
An exposure ontology is used to represent an exposure event—in this case, exposure of a zebrafish to Aldicarb in water during a laboratory study—which is integrated into a knowledge graph that combines data from multiple sources. Previous model organism studies have shown that the zebrafish gene, rpl11, is important for the microcephaly phenotype. The human ortholog is *RPL11*, enabling generation of hypotheses regarding the potential impacts of Aldicarb exposure in humans. Canine information can also be included, via its ortholog F6XQS3, but it will require development of standardized dog data sets.
SOURCE: Thessen presentation, December 2, 2021.

data resource can be used to improve pet health care, said Hughes, giving the example of anesthesia mortality, which is twice as high in cats as in dogs. To identify risk factors linked with mortality, researchers created over 300,000 structured anesthesia records from nearly 1 billion raw medical notes, which they are using to identify risk factors and build predictive algorithms that can provide real-time feedback to veterinarians to make anesthesia safer for cats, said Hughes.[74]

With the aim of improving preventive health care, Mars Petcare's Pet Insight Project uses the collar-mounted Whistle FIT activity tracker to monitor a wide range of behaviors in dogs such as eating, drinking, chewing, and scratching.[75] Synchronizing thousands of hours of video footage from citizen scientists with 11 million days of their dogs' Whistle FIT data, researchers developed a deep-learning algorithm that can accurately classify dog behavior based on activity tracking alone (Chambers et al., 2021). Validation of the algorithm found that it correctly identified eating 94 percent of the time and

---

[74] See https://www.thetimes.co.uk/static/data-analytics-dogs-cats-ai-machine-learning-mars-petcare-raconteur (accessed January 20, 2022).

[75] See https://www.petinsight.com (accessed January 20, 2022).

drinking 98.8 percent of the time and could identify more nuanced behaviors, such as sniffing and licking, to a lesser extent. This activity tracker provides a new way to capture nuanced behaviors as a biomarker, said Hughes, enabling early detection of changes that could result from disease, such as increased scratching. Hughes also noted that Whistle GPS trackers could enable the study of social interactions in groups of dogs.

Hughes described several other applications of machine learning that were developed to analyze Mars Petcare's large data sets, including automated evaluation of chest and abdominal radiographs in dogs and cats; reliable, automated cancer grading of biopsy slides using mitotic figure analysis; and early prediction of feline chronic kidney disease (CKD). CKD is the main cause of death in cats over 5 years old and has been difficult to diagnose in its early stages, said Hughes. Researchers at Mars Petcare developed a predictive tool that uses a combination of six common cat health measurements to predict CKD up to 2 years before a traditional diagnosis could have been made, enabling care to potentially slow the progression of disease, said Hughes (Bradley et al., 2019).

Embark: Using Dog Genetic Data to Drive Translational Discovery

The domestication of dogs from wolves to phenotypically differentiated closed breeding populations has provided a model for studying rapid evolution, genetic load, and personalized genomics, said Adam Boyko (Embark Veterinary, Inc./Cornell University). Genetic studies of dogs, accelerated by the sequencing of the dog genome (Lindblad-Toh et al., 2005), have identified hundreds of likely causal variants underlying Mendelian traits, most of which are potential models for human disease.[76] Performing genome-wide association studies on 4,200 dogs, Boyko and colleagues identified new genes associated with six complex diseases and two morphological traits (Hayward et al., 2016).

Village dogs offer a window into early dog evolution because they have been largely preserved from the type of artificial selection used to generate closed breeding populations, said Boyko. By comparing linkage disequilibrium in dogs around the world, Boyko mapped the likely origin of domesticated dogs to central Asia (Shannon et al., 2015).

Despite the progress in identifying Mendelian and large-effect loci in dogs, there has been slower progress in elucidating the genetics underlying behavior, longevity, and complex diseases such as cancer, which are big-data questions and will require large-scale studies on hundreds of thousands of dogs, said Boyko. Noting that direct-to-consumer genetic testing of humans has built data sets on the order of tens of millions of individuals, Boyko developed a similar

---

[76] See https://www.omia.org/home (accessed January 20, 2022).

model for dogs. Embark[77] was begun in 2015 and now has over 900,000 dogs in its database, each of which has been genotyped at over 230,000 markers, making it probably one of the biggest dog genetic databases in the world, said Boyko. The database consists of roughly 70 percent mixed-breed dogs, 30 percent purebred, with the remainder wolf dogs and village dogs. Boyko and colleagues are using the data for discovery and have identified novel loci associated with blue eyes in Siberian huskies (Deane-Coe et al., 2018) and with a mottled coat pattern (roaning) in Australian cattle dogs (Kawakami et al., 2021); the causal mutation in each case is a DNA duplication.

To obtain health data, Embark surveys owners annually, with over 180,000 surveys completed in 2021. Owners rate their dogs' general health and report in detail on organ systems, symptoms, and diagnoses. The researchers found that inbreeding has a strong negative correlation with longevity, with outbred and mixed-breed dogs outliving inbred and purebred dogs at every body size (Yordy et al., 2020). Noting that big data is less clean than small data, Boyko aims to maximize statistical power by balancing sample size and the signal-to-noise ratio, using automated quality control and tracking. Boyko has been particularly interested in automating discovery, which enables the researchers to explore associations across the entire genome and then follow up with a targeted validation. Million-dog cohorts are too large for standard approaches like PLINK,[78] said Boyko. It is important to standardize not only data models but also how the significance of genetic variants is communicated to owners. This can vary with breed and is complicated for mixed breeds, he noted.

Boyko mentioned some limitations of using dogs as models for humans, including incomplete age data and differences in the degrees to which stature, reproductive history, smoking, and inbreeding influence health outcomes in the two species.

Data and Resource Sharing with the American Kennel Club

About eighteen percent of U.S. dogs are from litters registered with the AKC, said Mark Dunn (AKC). The AKC has 175,000 active breeders who volunteer 50,000 DNA specimens from pedigreed dogs with known familial history each year; the DNA collection now includes about 600,000 specimens going back over 20 years. The AKC also enrolls 70,000 spayed and neutered mixed-breed dogs each year. Dunn added that the AKC database was recently

---

[77] See https://embarkvet.com (accessed January 21, 2022).

[78] PLINK is a free, open-source whole genome association analysis toolset, designed to perform a range of basic, large-scale analyses in a computationally efficient manner. See https://www.cog-genomics.org/plink (accessed February 11, 2022).

augmented with third-party demographic profiles, which may be useful for correcting issues of low representation from underserved communities.

The AKC supports researchers through consulting on the U.S. dog population, particularly with respect to breed-specific population data (aided by its litter registry); this has included geospatial mapping to 10-digit zip codes and geolocation coordinates. The AKC uses its large social media presence and contact management capabilities to drive interest, awareness, and participation in research, directing dog owners toward clinical surveys or trials. The AKC recently transitioned its genotyping method from STRs (short tandem repeats) to SNPs (single nucleotide polymorphisms), increasing its analytical capabilities, and expects to double its DNA collections to 100,000/year. The AKC is also seeking to enhance its data collection on health outcomes and cause of death and is interested in collaborations to capture phenotypes and combine them with genotype and pedigree data. The AKC has a strong interest in the genetics of behavior and is working with James Serpell's group at the University of Pennsylvania to make the Canine Behavioral Assessment & Research Questionnaire (C-BARQ)[79] canine behavioral evaluation tool available to all AKC members. Dunn reached out to the researchers for input in building infrastructure for sharing the AKC's sophisticated data collection, noting that purebred AKC dogs have the advantage of known pedigrees.

Dunn and participants from the research community considered how best to identify the most important diseases in each breed for investigation of possible environmental contributions. Over the last 5 years, the AKC has worked with its national breed clubs (called "parent clubs") to clarify each breed's most significant health concerns. This has resulted in over 180 letters detailing breed-specific health screening recommendations, which are publicly available in the "health" section of each breed page on the AKC website, said Dunn.[80] In addition, breeders who participate in the Breeders of Merit or Bred with H.E.A.R.T. programs are required to perform all the health testing required by their club.

Discussion: Accelerating Cross-Species Comparisons

Integrating Dog and Human Studies

Participants considered the feasibility of obtaining parallel data on pets and their owners by either recruiting the pets of participants in human stud-

---

[79] See https://vetapps.vet.upenn.edu/cbarq (accessed January 24, 2022).
[80] See, for example, http://cdn.akc.org/Marketplace/Health-Statement/Collie.pdf (accessed January 24, 2022).

ies or vice versa. Freya Mowat (University of Wisconsin–Madison) noted the expense of contacting human research participants for participation in an add-on pet study, when half do not even own pets. She suggested, at the very least, that every human study participant be asked whether and what type of pet they own. "Do you have a pet?" is an important question for any human longitudinal health study, given the increasing understanding of the impact of the human–animal bond on people's health, said Hughes. Promislow, who has addressed this question with leadership at the Baltimore Longitudinal Study of Aging[81] and the Framingham Heart Study,[82] commented that there is a high bar to adding even one more question to the surveys. He suggested bringing together leaders of human and pet prospective studies in a roundtable to figure out how to work collaboratively and support one another's research. Members of the dog community who can best identify the important non-monogenic breed-specific diseases need to sit down with their human-disease counterparts to pinpoint areas for collaboration, said Mowat. On the dog side this could involve the AKC parent clubs, said Dunn.

About 17 veterinary schools are currently affiliated with clinical and translational science awards (CTSA) to medical schools, which are designed to train clinicians in translational research, and the NIH has several initiatives aimed at bringing MDs and DVMs together around shared diseases, said Trepanier. Many of the affiliated medical schools have prospective longitudinal studies in humans, and work is under way to link veterinary studies to those human studies. However, in only six of these schools are the veterinary and medical schools housed on the same campus.

Participants were asked how they envision pairing human and pet data. Boyko suggested doing paired GWAS of dogs and their owners, to see which phenotypes map to the dog genome and which map to the human genome. Castelhano is currently integrating data from human and dog biobanks, with bioinformaticians working to standardize the platforms. The ICDC is funded by the NCI's Cancer Moonshot Initiative,[83] which supports data sharing and harmonization, noted LeBlanc. Once the Moonshot-funded Center for Cancer Data Harmonization[84] finishes creating its harmonized data model, ICDC data will be combined with human data in the Data Commons, amplifying the importance of uploading data to the ICDC, added Thessen. Although the ICDC was initially conceived as a repository of canine cancer genome data,

---

[81] See https://www.blsa.nih.gov (accessed January 24, 2022).

[82] See https://framinghamheartstudy.org (accessed January 24, 2022).

[83] See https://www.cancer.gov/research/key-initiatives/moonshot-cancer-initiative (accessed January 25, 2022).

[84] See https://datascience.cancer.gov/data-commons/center-cancer-data-harmonization-ccdh (accessed January 25, 2022).

it has broadened to include various types of data related to cancer development that have a rationale for comparison to humans, and this offers a huge opportunity for inclusion of environmental exposures, added LeBlanc. The ICDC has been opened for 6 months and contains data on bladder cancer, osteosarcoma, and glioma, including genomics, transcriptomics, and imaging, said LeBlanc, who encouraged researchers to use it.

Citizen Science: Opportunities and Concerns

This is a "golden age" for citizen science, said Boyko, highlighting Darwin's Ark[85] and Loyal for Dogs[86] in addition to Embark, the Golden Retriever Lifetime Study, and the Dog Aging Project. There is currently no comprehensive list of these projects, but the American Veterinary Medical Association (AVMA) has a clinical trial database akin to clinicaltrials.gov, said LeBlanc, and it would be straightforward to register studies on that website. Hughes suggested that the AKC publish a list.

Citizen science raises multiple concerns regarding data health, noted Dunn, with owners frequently misreporting phenotypes and breed status. Ostrander and LeBlanc noted the importance of verifying breed status genetically. Health outcome data need to be collected in a way that ensures it is all correct, or that enables drilling down to the useful data, said Dunn.

Dunn has also seen big differences in reported household demographics between subjective and third-party (objective) reports. There is also selection bias; the people most likely to work with researchers are the ones already in the database. Dunn also noted ethical concerns surrounding informed consent, privacy, and commercialization of data (see section on Ethical Considerations of Using Companion Animals as Sentinels).

Sharing Commercially Held and Biobanked Data

Shelton asked about open access for academic and other researchers to the huge databases of Embark and Mars Petcare. Boyko described three ways in which Embark makes its data available: Raw data for published papers are submitted to public archives, researchers can contact Embark directly to obtain consent for the data they need, or researchers can obtain data directly from pet owners. This last method worked "spectacularly well," said Ostrander, when Embark contacted owners of pets with a particular disease and invited them to get in touch with Ostrander. Mars Petcare's Science Engine integrates data

---

[85] See https://darwinsark.org (accessed January 24, 2022).
[86] See https://loyalfordogs.com/about (accessed January 24, 2022).

from across the business, said Hughes, and researchers are invited to partner with its Data and Analytics team. In addition to various institutional partners who work with its data, Mars Petcare collaborates on clinical trials with VCA Animal Hospitals,[87] BluePearl Pet Hospitals,[88] and others, added Hughes.

The AKC works closely with the Orthopedic Foundation for Animals (OFA),[89] whose Canine Health Information Center (CHIC) program[90] database is an actively collecting biobank that includes scored X-rays, health testing, and health outcomes for thousands of dogs, said Dunn. Many breeders who do Embark testing submit these data to the CHIC, added Boyko, which could enable cross-referencing of hip X-rays with genotype data. Castelhano noted that the CHIC is among a network of biobanks around the country; researchers can contact any biobank in the network and will be referred to whichever one contains the specimens they need.

Purebred versus Crossbred Dogs as Models for Environmentally Induced Cancers

Participants considered the relative advantages and disadvantages of using purebred versus mixed-breed dogs to study the influence of environmental agents on the development of cancer. Dye raised the question of whether, in a breed at high risk for a particular cancer, the hardwired genetic risk might potentially overwhelm the ability to detect the influence of an environmental exposure—or on the other hand, in cases where gene–environment interaction is critical, would that breed demonstrate particular sensitivity to the environmental agent? That depends on the genetic architecture underlying the risk, said Boyko. As noted by Page and Trepanier, the importance of the environment to cancer incidence in susceptible purebred dogs has been demonstrated: For example, pesticide exposure increases the risk of bladder cancer in Scottish Terriers, a high-risk breed, whereas ingestion of vegetables reduces the risk of bladder cancer in both people and dogs and improves survival time in patients with the disease (Glickman et al., 2004; Raghavan et al., 2005; Tang et al., 2010). Page also added the GRLS finding that 50 percent of cancer deaths were due to hemangiosarcoma, which is uncommon in dogs outside the United States and thus suggests the presence of an environmental influence. However, if the cancer susceptibility is fixed within a breed, as is suspected for the Scottish Terrier, it would not be possible to elucidate the underlying genetics by studying that breed alone; dogs outcrossed from the sensitive breed would be used to

---

[87] See https://vcahospitals.com (accessed January 24, 2022).
[88] See https://bluepearlvet.com (accessed January 24, 2022).
[89] See https://www.ofa.org (accessed January 24, 2022).
[90] See https://www.ofa.org/about/chic-program (accessed January 24, 2022).

identify the region(s) of the genome responsible for the predisposition to risk, said Boyko. Chromosomal analysis of outbred dogs narrows down the region of interest and allows for a much finer GWAS, added Ostrander.

High-cancer breeds may have certain additional factors that sensitize them to an environmental exposure, said Boyko, citing a study from the Ostrander lab demonstrating the importance of genetic background in determining whether a cancer-predisposing mutation in the Kit ligand gene was expressed in poodles. Although the Kit ligand mutation was found in both black and white poodles, only poodles carrying a particular allele in another gene, which was associated with black coat color, expressed it (Karyadi et al., 2013). Trepanier added that GWAS studies in dogs have failed to identify many cancer drivers, suggesting that most of these tumors require a gene–environment interaction. Boyko, Trepanier, Dye, and Ostrander concurred that purebred and mixed-breed dogs would reveal different things depending on the genetic architecture of risk, and it was important to study both.

Incorporating Environmental Measures into the Database

Noting that chronic noncommunicable diseases in humans are associated with environmental factors acting on multiple genes, with each gene contributing a relatively small amount to the health effect, Birnbaum wondered how environmental measures would be incorporated into these databases. The environment, said Boyko, can be thought of as just another phenotype, albeit a highly complicated one. This involves collecting data over a time series—not just for exposures but for phenotypes like obesity, which is not a single data point, and which can be combined with environmental inputs like nutrition and exercise over time. Time-series data may be available from various sources—for example, the genetics committee of the American College of Veterinary Ophthalmologists (ACVO) reviews annual screens for potentially heritable disease in dog eyes, noted Mowat.

Jones encouraged researchers to leverage existing, geographically linked exposure data as an additional resource to supplement the collection of environmental and biospecimens, noting that "those data exist and have been mined and cleaned and resourced for lots of different purposes" and could function as a hazard identification tool, particularly for investigating new exposures. LeBlanc has considered geospatially mapping the residences of dogs enrolled in osteosarcoma drug trials. The veterinary schools where these dogs are treated are geographically dispersed across the United States and Canada, and LeBlanc has observed differences in outcome by location. This could reflect differences in care or environment, she noted, cautioning that with only 350 dogs in the cohort, these finding were not statistically significant.

Birnbaum suggested that researchers explore the NIH All of Us research program,[91] which aims to recruit a million volunteers around the United States and is now in its second phase and collecting basic environmental data, possibly including a question about pet ownership. This will be a huge study with follow-up and opportunities to access the data, she added. A participant noted that the 2000 U.S. Census included a question about whether there was a dog in the household, but this was eliminated from the 2010 and 2020 questionnaires. Restoring that question could help address the denominator problem, which is the lack of knowledge regarding how many animals reside in a given location, said Jones. The pet industry and the AKC have both lobbied to restore the question, added Dunn. The AVMA could join that conversation, suggested LeBlanc.

EQUITY, ETHICS, AND POLICY

Participants discussed issues of ethics and equity raised by the use of companion animals in research. Speaker presentations considered the opportunities to proactively ensure that these research methods do not perpetuate structural inequities or discriminate against disadvantaged populations; the appropriate role of pet owners in these studies; and the collaboration opportunities (e.g., multi-sectoral collaboration mechanisms, community engagement strategies, educational opportunities) needed to effectively implement the core capacities and interventions of One Health principles within communities.

Ethical Considerations of Using Companion Animals as Sentinels: Research Subject Protections, Citizen Science Issues, and Shared Health

Embedding ethical considerations into research strengthens study design and validity, and is the right thing to do, said Lisa Moses (Harvard Medical School), who is one of two ethicists consulting on the Dog Aging Project. "Everyday ethics," said Moses, requires examining projects to identify issues like cognitive bias or hidden clashes of values, "those common issues that trigger vague feelings of discomfort or red flags . . . should we pay enrollees, or why is the project having trouble enrolling diverse participants?" Applying basic principles of ethical analysis when planning projects, such as listing everyone who may be affected by a project's activity or outcome, can help clarify these issues, she said. When planning partnerships with prospective commercial partners or private funding sources, each prospective partner's objectives should be compared to prevent future conflicts "before the challenges emerge."

---

[91] See https://allofus.nih.gov (accessed January 25, 2022).

Though significant human–animal relationships have existed for millennia, the meaning of these relationships has changed as humans have urbanized and become increasingly likely to live alone and to delay marriage and childbearing, said Moses. Pets now fill holes in the support network for millions of people. By and large, ethicists consider companion animals as having a special, protected moral status compared to other animals (Yeates and Savulescu, 2017). In Moses's view, the primary argument supporting this special status is relational, where the complexity of human–pet relations resembles those between humans, and thus lends them moral weight (Ashall and Hobson-West, 2017).

Though this workshop mostly covered observational studies, which carry less potential for harm than interventional studies, Moses highlighted three basic ethical issues that remain relevant:

1. *Informed consent does not exist for animals.* Whereas human subjects give informed consent to protect their own autonomy, consent by a pet owner reflects their ability to preserve the value (economic, emotional, etc.) the animal has for them and does not require prioritizing the best interest of the animal (Ashall et al., 2018; Gray et al., 2018).
2. *Pets are not protected by laboratory animal research regulations.* Although dogs and cats are protected species under the Animal Welfare Act, most of what is assessed by an Institutional Animal Care and Use Committee (IACUC) does not apply to animals housed outside of labs (Walker and Fisher, 2018). Moses urged researchers to seek additional review of their study plans by a Veterinary Clinical Trial Ethical Review Committee (Bertout et al., 2021).
3. *Payment or other compensation for enrollment can be coercive when pets are the subject, due to inequities in access to veterinary care in the United States.* If the compensation is veterinary care, that might be the only care the pet receives, in which case it incurs a moral obligation on the investigators. Another concern is therapeutic misconception, where owners believe mistakenly that the research will directly benefit their pet.

While citizen science has tremendous potential to democratize science, it carries its own ethical concerns, added Moses. It is essential to incorporate the principles of reciprocity, where participants benefit from their involvement, and reflexivity, where investigators examine the influence of their own beliefs and practices on the research. Other useful strategies include diversifying participants, which improves data quality, and building in learning objectives to enhance participants' scientific literacy. There are well-developed tools for engaging in these partnerships, including a NASEM consensus study report, said Moses (NASEM, 2018; Soleri et al., 2016).

The Potential for Coercion in Interventional Studies

The perception of coercion in interventional studies was of particular concern to participants. Nicole Erhart (Colorado State University) noted the concept that compensation could be coercive when it covers standard-of-care treatment. To offset the perception among pet owners that this is the only way they can afford to care for their pets, researchers could ensure that participants have another way to obtain care, which may involve steering them toward other resources, said Moses. Another strategy is for a social worker to privately discuss access to pet care with the owner during the original consent process. Trepanier raised the scenario of a cancer clinical trial that enrolls patients through a free veterinary clinic, with compensation in the form of specialized care, such as free tumor biopsy, genetic analysis, and therapy. This presents an ethical issue, said Moses, if the only way people can obtain that level of care is by participating in the study. "The potential for therapeutic misconception and misidentification is huge when people are desperate," she added. When they cannot obtain the same level of care elsewhere, it is essential that both owners and researchers are aware of this and that owners are clearly informed of their obligation to their pet and their ability to withdraw at any time. For complex situations like these, Moses suggested writing consent documents that address all the essential ethical concerns on the first page, with more details in other parts of the document. A clinic offers the opportunity for face-to-face communication, which goes a long way toward finding the participant's source of discomfort or conflict. "When it becomes the unspoken thing, that is when things get dangerous, and people do not feel comfortable speaking up when they are not sure . . . what the implications of their decisions are," she added.

Moses also expressed concern with the question of euthanasia, particularly in oncology studies, where owners want to contribute to the science on the one hand but also want to decide when to end their pet's suffering on the other, and recommended involving social workers in that discussion.

Veterinary cancer patients resemble pediatric cancer patients whose parents give consent for their participation in clinical trials, suggested Page. In both cases, problems arise from a lack of full medical coverage and the potential coercion embedded in free care. The main difference between veterinarians and pediatricians, said Moses, is that veterinary patients are a different species with different needs than humans, and veterinarians understand this in a way that owners often do not. Parents can usually be assumed to know their child best and advocate for his or her best interest. In contrast, veterinarians are called on to advocate for and protect their patients in a different way, said Moses.

One Health Approaches in Arctic Indigenous Communities

Persistent Organic Pollutants: Contamination of Arctic Wildlife and People

The rise in synthetic chemical pollution is outpacing all other measured agents of global change, including habitat loss and the increase in atmospheric levels of CO_2, but synthetic chemicals remain the least studied of all these agents (Bernhardt et al., 2017). The heat of equatorial climates causes light- to intermediate-weight pollutants to evaporate and disperse in the earth's atmosphere, where they condense and are deposited in colder regions. Over time, through cycles of volatilization and deposition, persistent organic pollutants (POPs) work their way to the highest latitudes, and this has made the Arctic the most polluted place on earth for these compounds, said Frank A. von Hippel (University of Arizona). POPs bioaccumulate in animals and magnify in food webs, with long-lived predators such as polar bears and whales registering the highest concentrations of POPs on earth. These chemicals are further concentrated in the Arctic indigenous people who subsist on these animals, said von Hippel, adding that mining, oil extraction, and about 600 former defense sites contribute to the pollution in Alaska (von Hippel et al., 2016).[92]

Cancer is the main cause of death among Alaskan natives, with significantly higher rates of lung cancer and colorectal cancer in Alaskan native men and women compared to the general U.S. population.[93] Additional health patterns among indigenous Alaskans that may relate to POP exposure include thyroid disease, diabetes, heart disease, birth defects, and other reproductive health problems.

Sled Dogs as Sentinels for Environmental Exposures

Dogs are a major part of Alaskan culture, where they are relied on for transportation, subsistence hunting, safety on the ice, and protection from predators, in addition to sport, said von Hippel, noting that dogsled races form the centerpiece of cultural events and that the Iditarod race commemorating the 1925 diphtheria antitoxin run to Nome celebrates "one of the first One Health projects." Many Alaskans also rely on their dogs for emotional health and sobriety. Arctic dogs are more like workers than pets, with paral-

---

[92] See https://www.niehs.nih.gov/research/supported/translational/peph/assets/docs/the_use_of_sentinel_species_in_health_disparities_research_508.pdf (accessed January 27, 2022).

[93] See http://anthctoday.org/epicenter/publications/Cancer_50year_Report/antr_fifty_year_report_web.pdf (accessed January 27, 2022).

lels to military dogs or rescue dogs, which may raise different ethical issues, noted Birnbaum.

Alaska presents unique issues for studying companion dogs as sentinels of environmental exposure, said von Hippel. Most dogs live outdoors, so they are not a suitable model for studying household exposures. Arctic communities are very remote, and transportation is mostly by small plane, and there are no facilities or housing for researchers. These regions have high levels of poverty and virtually no veterinary care, which amplifies ethical concerns. The Arctic has a broad host spectrum for rabies, and rabid dogs pose a major public health risk. Dogs can also be a source of conflict when they get caught in trap lines. On the other hand, Alaskan dogs and people eat the same food, making dogs a good sentinel for POPs exposure, which primarily occurs through the diet; sled dogs have the same profile of contaminants as their owners and suffer similar health effects, said von Hippel. The research enterprise itself can bring improvements to the community in terms of increased access to veterinary care and jobs, improved infrastructure, improvements to human health through treatment of zoonotic diseases, and conflict reduction through community engagement. Although many communities are legally dry, drug and alcohol addiction is a major problem throughout the Arctic, and community engagement in research can help address this problem by giving people agency and improving resilience, he added. The remoteness of these communities limits confounding variables, and there is excellent scientific literature on the use of sled dogs as sentinels throughout the Arctic.

Community-Engaged Research to Study Exposures in Alaskan Natives and Dogs

Von Hippel's research always originates with a request for help from the community, generally out of concern for contamination of their villages and subsistence foods. For example, a health aide on St. Lawrence Island, concerned about whether the high rates of cancer in her community might be caused in part by exposure to thousands of barrels of unknown contaminants that had been left behind by the U.S. military, contacted a nongovernmental organization (NGO) that recruited scientists to work on the problem. Another tribe asked for von Hippel's help dealing with pollution from World War II defense sites in the Aleutian Islands. The community originates the question, and a working group forms between the scientists and community members to strategize a way to get an answer. Community-engaged research is essential to avoid perpetuating inequities or discriminating against disadvantaged populations, and ideally takes the form of community-based participatory research (CBPR), said von Hippel (Wallerstein and Duran, 2010). When working with indigenous populations, it is essential to obtain consent from the community and tribe in addition to the dog's owner, and the community should benefit from the work, he said. The tribe owns von Hippel's data and he reports results to them before publishing. More researchers are needed in the Arctic and there are opportunities for collaboration, said von Hippel, though he cautioned that this work proceeds on a long timeline, which requires that scientists educate funding agencies and build a trusting relationship with the community before writing the grant proposal.

There are important parallels between climate change and synthetic chemical pollution in terms of their effect on the Arctic, added von Hippel. Like POPs, the effects of climate change are most pronounced in the Arctic, which is warming at twice the rate of the rest of the planet. Both POPs contamination and climate change are cases of environmental injustice, exerting their greatest effects on populations that did not create or use these compounds. Declines in polar bears and other Arctic wildlife are due more to POPs exposure than to climate change, and the effects of the two exacerbate one another: Climate change increases the geographic spread of disease while POPs compromise immune function and amplify disease susceptibility. Noting that the attention received by synthetic chemical pollution is disproportionately low relative to the increasing urgency of the problem, von Hippel cited a recent paper advocating for an intergovernmental panel to tackle global chemicals and waste (Wang et al., 2021).

Aligning Health Care for a Bonded Family Society

Two-thirds of U.S. households have pets—considerably more than have children—and as many as 95 percent of these consider their pets to be family members.[94] Many families with strong human–animal bonds, or "bonded families," have limited assets; these include individuals employed in the service of others or retired on limited incomes, with a disproportionate representation of minorities and women, said Michael Blackwell (University of Tennessee, Knoxville), citing a study by the Access to Veterinary Care Coalition (AVCC).[95] Those with the least often demonstrate the strongest human–animal bonds, and this includes people who are unsheltered—in some cases because available housing options exclude their pet. Twenty-eight percent of pet owners in the AVCC study reported a barrier to veterinary care in the pre-

---

[94] See https://www.prnewswire.com/news-releases/more-than-ever-pets-are-members-of-the-family-300114501.html (accessed May 23, 2022).

[95] See https://pphe.utk.edu/wp-content/uploads/2020/09/avcc-report.pdf (accessed January 27, 2022).

vious 2 years, said Blackwell. These barriers to care were not just financial but included practical hurdles like transportation. Pets exist throughout society because they benefit humans, both physiologically and psychologically, and this results in reduced morbidity and mortality (Matchock, 2015); therefore, rather than trying to remove pets from owners who cannot afford veterinary care, Blackwell wants to change the system to help these families.

Societal support for pet care is also becoming a public health imperative, argued Blackwell, as climate change broadens the range of disease vectors and leads to an increase in zoonotic diseases. The incidences of both ehrlichiosis and Rocky Mountain spotted fever have risen dramatically in recent years.[96] Ignoring the health care needs of bonded families would adversely impact the health of the nation overall, said Blackwell. But addressing this requires revisiting public policies, such as those that concern housing, and factoring in the needs of bonded families. This requires a paradigm shift to One Health, which recognizes that efforts to improve health outcomes for humans must take the health of animals and the environment into account, said Blackwell.

There was also discussion about the particular needs and potential pitfalls of conducting research in which bonded families participate. "In my experience, pets very often will reflect the reality of the family," said Blackwell. For example, low-income families are more likely to feed their pet the same food they eat rather than buying pet food. This can lead to similar health conditions in people and their pets and it could be beneficial to study conditions like diabetes or obesity in the companion animals as well as the people. Treating the bonded family as a unit would enable researchers to engage the companion animal in studies of family dynamics around mental and emotional health in addition to infectious disease.

Discussion: Equity, Ethics, and Policy

Different Uses of Companion Animals Lead to Different Shared Exposures

Shelton was intrigued by the shared diet between sled dogs and humans in their community and its ability to shed light on the physiological effects of contaminants. Von Hippel agreed, citing research in Greenland, Canada, and Alaska on persistent organic pollutants and associated disease states, which compared sled dogs with Arctic foxes, polar bears, and their human owners; "the parallels are really striking," he said.

---

[96] See https://www.cdc.gov/ehrlichiosis/stats/index.html and https://www.cdc.gov/rmsf/stats/index.html (both accessed January 27, 2022).

Unlike Arctic sled dogs, military dogs live and sleep with their individual human comrades-in-arms, so exposures to stressors may be even more linked in these parings than with pets, noted Paula Genik (Colorado State University). Responding to Blackwell's comment that the emotional, psychological, and mental health of both dog owners and their pet dogs are deeply affected by their common lived experience, Genik suggested that military dogs are excellent sentinels of combatant mental health, and that perhaps Arctic dogs could be able to provide insight into the mental health of their owners. Von Hippel agreed that relationships between owners and working dogs are fundamentally different from those between owners and pets and suggested that researchers explore how to better incorporate working dogs into the One Health framework.

Research, Service, and Education: Balancing Priorities

Sue VandeWoude (Colorado State University) expressed concern about how to prioritize the need to provide underserved communities with human and/or veterinary medical care, as well as research needs. Disease distribution and trends among companion animals in these communities is poorly understood, said Blackwell, so that has to be studied in order to determine what kind of medical care to prioritize. Based on its survey of pet owners and veterinarians, the AVCC report recommended development of new service delivery models, including methods for controlling costs, to find the most economical way to provide veterinary services when income is limited.

Promislow asked if there is an ethical approach to using compensation to attract a more diverse participant pool to observational studies, particularly in communities that might not have easy access to veterinary care. In the case of a disadvantaged community, you can find more creative solutions, said Moses, such as working with local care providers to offer a free or low-cost clinic event that supports the community as a whole, while building trust in the research project. The key with citizen science is building relationships within the community and asking participants about what help they need, she added; a project that addresses systemic barriers to care in the community would significantly help with participant diversity and equity.

Families need help understanding the needs of their nonhuman family members, and with so many people turning to the internet for advice, pet owners need guidance on how to use research tools to obtain reliable information, said Blackwell. There are many ways to incorporate education into citizen science, said Moses, and it is important to help participants increase their scientific literacy. This might include training in data collection methods, which is important for study validity, as well as directed training for adult learners. For large-scale projects that involve many communities, Moses suggested partnering with local schoolteachers, who are always looking for ways to engage their students with science and incorporating a trained educator into the project design team. "[It] is really about maximizing very explicit and direct educational opportunities for participants," said Moses.

Cultural Competence and Establishing Partnerships in Underserved Communities

When working with underserved communities, communication can be a challenge, said Tracy Webb (Colorado State University), noting that not everyone has reliable access to email or participates in social media, and there may be barriers to what individuals are willing to report. Citing von Hippel's emphasis on both top-down and bottom-up communication, with the community directing the research to address the questions most important to them, Webb asked how to actualize these concepts and establish effective two-way communication in underserved communities.

In Arctic communities, and indigenous communities in general, "there is no substitute for being there," said von Hippel, so "the [funding] agencies have to be willing to fund travel to get to these remote places." Von Hippel's group makes many trips and does community presentations, which involve local leadership, and these have to be done in person because most areas do not have broadband internet. In Australia, where he works with an aboriginal community, von Hippel reaches the community through a local radio show. "We talk to the radio host about the research and what we are finding. So it's a matter of finding the right means within each local community that reaches people."

It is critically important to gain cultural competencies around the cultures one is studying, said Blackwell. Historical experiences, such as the Tuskegee Syphilis study,[97] have molded perceptions of research within these communities. Often the only connection they have with outsiders around their pets is through animal control, which translates as policing, so we need to be attentive to those perceptions, said Blackwell. One example that works, said Moses, is a program called Pets for Life,[98] whose workers go door to door, ask whether there are pets in the home, ask what the residents need to maintain the pet, and provide it. Partnering with an organization like this, which has established trusting relationships in the neighborhood, is the most direct way to begin a research project, suggested Moses.

---

[97] See https://www.cdc.gov/tuskegee/timeline.htm (accessed February 27, 2022).
[98] See https://humanepro.org/programs/pets-for-life (accessed February 11, 2022).

Research with a long timeline can be challenging, said Blackwell, but underserved and underresourced communities may require a long period of engagement to establish trust with researchers. Noting that Promislow took 3 years to establish infrastructure for the Dog Aging Project, Don Ostrow (educator, NYC) asked how to establish research models that acknowledge the necessary time to establish trust in the community, challenge basic assumptions within the research community, and remove obstacles to providing genuine two-way communication.

Von Hippel writes memoranda of understanding (MOUs) between the university and the tribe that clearly state the relationship, data sovereignty, and other issues important for community participation. "I think that protection for the tribe is critical so they feel confident that they are not being exploited and that the work is really to benefit the community, rather than just researchers coming in, doing their work, and never hearing from them again, which is the way it used to be that was so problematic," he added.

Moses suggested involving an expert in research ethics in the initial planning phase, which can speed up the process. Blackwell suggested engaging individuals with leadership roles, which may vary according to the community. For example, within the Black community this has traditionally included ministers and teachers, and Blackwell suggested starting by building relationships with these individuals, adding that ". . . if one is serious about this, from the standpoint of respecting me as a minority and recognizing that there is some history that I embody and has been handed down, that literally defines the culture—if you can't come in respecting my culture, you will be obvious from the beginning by what you say, what you don't say, and how you go about doing it."

IDENTIFYING RESEARCH GAPS AND SETTING A RESEARCH AGENDA: EXPLORING NEXT STEPS FOR THE PATH FORWARD

Planning committee members summarized facilitated discussion tables in which workshop participants considered lessons learned from the workshop and strategies for charting a path forward. This was followed by open discussion with suggestions to move the field forward from planning committee members, invited speakers, and audience members.

Obtaining Data on Pets and Exposures

Demographic information is an important denominator for some of this research and is lacking for all populations, including purebred dogs, mixed breeds, and cats. The AKC, AVMA, and industry can help obtain this information, said William Farland (Colorado State University, emeritus); adding a question about pets to the 2030 Census would also be useful. One approach to improving the denominator, suggested Promislow, would be if a team of human demographers and geospatial scientists worked with the databases from Mars Petcare's veterinary hospitals and clinics to develop models that could derive estimates of the denominators based on existing information such as breed, age, and size. This is important for understanding risk factors and incidence of disease—particularly for rare diseases, said Promislow, which may not be so rare in certain populations.

Danielle Carlin (NIEHS) suggested developing studies that would allow physicians and veterinarians to collect biospecimens from humans and pets simultaneously—for example, through existing mobile clinics. Promislow noted that dog owners are often more willing to do things for their dog than for themselves, which could extend to the collection of dust and other samples in the home. The fact that pet research would be included with the human research might motivate owners to engage more in this kind of work, and it underscores the value of piggybacking dog studies onto longitudinal human studies, he added, which would benefit both studies.

Interdisciplinary Training of Researchers, Physicians, and Veterinarians for One Health

Johannes discussed the need for physicians and veterinarians to be trained in public health, particularly as veterinarians become more specialized and 80 percent go into private practice, where their focus is on caring for patients. Most veterinary residents are trained at universities, and it is the universities' role to convey the importance of participating in One Health, he suggested. Ryan suggested encouraging veterinary students to take public health courses and, in some cases, to complete dual DVM and MPH degrees. The fields of oncology, internal medicine, and nutrition would dovetail well with public health, said Wakshlag, and DVM programs could be tailored to include an extra year to accommodate this training. The problem seems to be funding, rather than interest, he added.

Greater support is also needed for the teaching of comparative medicine and research in medical schools because this trained medical workforce will be crucial for collaborating on research with practicing veterinarians, said Carlin. These companion animal studies can be used to train interdisciplinary personnel in public health, medical, and veterinary schools, as well as professionals in the social sciences who will be needed to bridge cultures and work closely with participants in community-based participatory research, added Farland.

Focusing on the animal side, Trepanier noted that veterinary students whose main interest is in primary care receive no training in research. To remedy this, Trepanier plans to pilot a veterinary student training program with CTSA One Health Alliance, to build interest in participating in citizen science in the primary care setting, and to view this participation as important for both the students' professional development and the growth of the field.

The primary care veterinary community plays a critical role in helping to gather data for companion animal studies, noted Promislow, but primary care veterinarians often do not have the bandwidth to participate in studies, and there is no funding for their participation. Furthermore, because many primary care veterinarians do not encounter the scientific literature in their training, they may not appreciate the value of doing this extra work for science. This presents an education challenge for researchers, who need to convince the veterinarians that it is worth their time to participate in a scientific study and to help them guide their clients through informed decision making about their own participation.

Practicing veterinarians could be engaged through continuing education (CE), particularly if it is provided in an American Association of Veterinary State Boards' (AAVSB's) Registry of Approved Continuing Education (RACE)[99] board-approved format that meets CE requirements for licensure, said Marguerite Pappaioanou (University of Washington). Community research education would probably be a very popular topic, added Page, who suggested building it into conferences and CE expectations for licensure. Veterinarians on the Dog Aging Project team are creating CE content as a way of "giving back" to the many practicing veterinarians helping with the rapamycin trial (TRIAD), said Promislow. Any veterinarian who participates in the DAP is offered CE, and once RACE approval is obtained, CE will be offered as distance learning on-demand, added Ruple.

Veterinarians are among the most respected sources of information to the general public, noted Farland, and it is important for researchers to collaborate and utilize veterinarians as a trusted resource. The research community could reach veterinarians through respected sources, such as the AKC and similar groups, the Toxicology Education Foundation,[100] and presentations at society meetings, he added. The American College of Veterinary Preventive Medicine would be a valuable partner, added Pappaioanou.

Translating the Science between Pets and People

How exposure data might be translated between humans and animals remains to be determined, said Carlin. She offered the example of physi-

---

[99] See https://www.aavsb.org/ce-services/race/program-overview (accessed February 2, 2022).

[100] See https://toxedfoundation.org (accessed Feburary 9, 2022).

ologically based pharmacokinetic (PBPK) modeling (Jones and Rowland-Yeo, 2013), which can help predict tissue-specific exposures to a drug in one population based on extrapolation from another population along with relevant physiological parameters. In this case, the pharmacokinetic (PK) for each species needs to be determined before effective comparisons can be made between species, said Carlin.

"This is about more than just genes and environment," added Farland, noting the observed roles of metabolism, receptors, the microbiome, and other aspects of the organism that modulate its response to environmental stimuli. Models could incorporate these other types of information, perhaps employing the toxicological concept of adverse outcome pathways, said Farland.

Data Integration

What does it mean to create compatible, harmonized data across species for so many different data types? There are many definitions of data and many components, noted Page. This may include questionnaire data and EMR data, but also sample data, which could take numerous forms, such as histopathology, which would require digitization. In addition, he asked, "Who is going to pay for that?"

Carlin emphasized support for a publicly available database and biobanks. Promislow envisioned the creation of a DNA Data Commons for dogs, where all the studies collecting DNA data would share information. For example, if a participant consents to having their dog enter an oncology clinic for treatment or a research study, the principal investigator (PI) or clinician would be able to access that dog's genetic data and analyze it for high-risk markers. Sharing DNA data brings challenges regarding consent, privacy, and consistency, with different platforms providing different information, but there may be easy ways to harmonize it, said Promislow. Genik cited a life sciences data archive maintained by NASA, which holds samples and '-omics data collected from astronauts and biological organisms both flown in space and on the ground and suggested that a similar structure might be useful to the NIH and the NIEHS. "I think what that comment underscores," said Promislow, "is that there are a lot of people who are not at this meeting who don't think about the questions that we're thinking about but who have solved some of the problems that we're thinking about. And one of our action items should be to think broadly and deeply about who else is out there solving different problems but with the kinds of solutions that we also might need to implement."

"We should not have to reinvent the wheel," said Page, suggesting instead that researchers expand the NCI's ICDC to contain exposure data and other types of data that would be generated from studies on a broader scale across species, with the vision of a common exposure or common genomic analysis arrangement. This will be expensive, added Page. There needs to be active work to promote data accuracy, collation, integration, curation, and accessibility; and that will require targeted funding, said Breen. The Dog Aging Project and Golden Retriever Lifetime Study cannot fund all the analytics needed to study the data they produce; rather, "they are an amazing resource for other people to adopt and fund," said Breen.

Data Collection: Long-Term Investment

There is an opportunity to build data infrastructure from scratch, but existing systems, such as human cancer registries or reportable illnesses in humans and pets, can also be leveraged for this effort, said Deziel. COVID-19 provides a case study of how a health data resource can be built quickly, given the will and the funding, he added. An EMR system for pets would help significantly, but the importance of this—and how it would benefit veterinarians and their patients—needs to be communicated to veterinarians, who are already overworked, said Deziel. Nonetheless, Promislow noted, there are also challenges getting data out of EMRs, as well as challenges getting veterinarians to share data.

Sample collection is expensive, making it a challenge to fund, particularly for larger-scale studies, said Promislow. Small clinical trials are straightforward, but for human studies and larger dog studies, with thousands or tens of thousands of specimens, the cost can be prohibitive. The research community needs to develop funding strategies, said Carlin, noting tremendous variability in access to resources.

Data Collection: Tapping into Existing Cohorts (Short Term)

There is a need to better understand exposure and the environment that companion animals live in, and this requires augmenting the information that veterinarians typically collect, said Farland, who reviewed methods covered at the meeting, including GIS, proximity to sources, and both active and passive sampling. He suggested leveraging existing environmental data and exposure information that has been collected by the NIEHS, the EPA (through its research and development program), and state and local health departments. Plenty of specimens have already been carefully collected and stored, noted Farland; the question is how to make the best use of these and surmount institutional barriers to making them available for study.

Deziel noted many opportunities for researchers to tap into existing cohorts, both human and animal. One example is the NCI-CONNECT cohort. A helpful resource would be a repository of existing cohorts that are amenable to additions. There are also underutilized data sets. For example, Poison Control[101] collects both human and animal data, as well as some location information. The research community needs to figure out how to collaborate with and access the data from Mars Petcare and other large data sources, added Carlin. This may include data from diverse sources, such as the Multi-Ethnic Study of Atherosclerosis (MESA).[102] NHANES collected data on water and dust samples and has access to Social Security and Medicaid data, as well as a significant amount of longitudinal data, said Carlin. Available GIS data could also be leveraged for studying companion animals, Carlin added.

Trepanier and Birnbaum discussed the possibility that the NIEHS and the NCI might provide supplemental funding to add pet biomonitoring and outcome data to existing longitudinal studies in humans. Studies initiated in disadvantaged communities might fit into other NIH institutes, including the National Institute on Minority Health and Health Disparities, added Birnbaum.

Continuing and Expanding the Conversation

Page suggested a follow-up meeting to consider what "data" really means and how to define it, and to create recommendations for building more robust data systems. Data convergence, pulling the dog and human data together, is another important conversation, said Shelton. Breen suggested establishment of focus groups "to develop the right questions to ask of these data."

"Where are sentinel animals poised to have the most impact?" asked Carlin, suggesting studies on physical health, exercise, and obesity, as well as mental, emotional, and social health and well-being. She also suggested setting up working groups to identify focus areas based on existing collaborative networks.

Shelton suggested that a communication committee could be tasked with getting information out to relevant stakeholders. Hughes also stressed the importance of having a central body to manage communications from the research community, with a consistent message. Breen said, "This is going to take a massive coordinated effort," which will require a partnership between scientists, veterinarians, companion animal owners, and others, as well as education of a new generation of leaders in research.

Community Research—Engagement and Equity

As researchers continue to grow the infrastructure for community science, they need to focus on making the science more accessible and understandable

---

[101] See https://www.poison.org (accessed February 11, 2022).
[102] See https://www.mesa-nhlbi.org (accessed February 11, 2022).

to the public, said Megan Dillon (North Carolina State University). This will enhance the value of the science and also generate support for funding, she added.

Science literacy and public awareness about these projects need to be increased, said Farland. The research community could work more with nonprofit organizations to reach out to communities, which could include working with animal shelters and rescue groups, and organizations that support people who are homeless, said Carlin, adding that researchers could engage marketing and community specialists to return results directly to the pet owners.

As this research community continues to engage in citizen science, it needs to put into practice community-based participatory research and find opportunities for two-way communication, said Farland. Promislow suggested changing the wording from "citizen science" to "community science," noting that some target participants who are not U.S. citizens may be put off by the former, whereas "community" embraces everyone. Promislow also cautioned that an equity issue arises in research where participants fund some of the cost, as only people with money can participate in these studies.

Noting that current leaders in companion animal research have been working with stakeholders for many years and have developed a strong level of trust, Breen suggested "a kind of legacy planning" to train the next generation of both researchers and community members.

CONCLUDING REMARKS AND POTENTIAL NEXT STEPS

The history of using animals as sentinels extends long before the canary in the coal mine to ancient times, said Birnbaum, when animals were used to determine whether a particular food was safe. Animals can be effective sentinels because of greater susceptibility, greater exposure, and shorter latency for adverse effects of environmental factors. Furthermore, domesticated dogs have been exposed to some of the same environmental selection pressures as people, and dog orthologous genes more closely resemble those of humans than do rodent genes.

This is a golden age of cancer research, said Birnbaum, with increasing awareness of the pressing relevance of environmental exposures. However, the latency of effects in humans makes it difficult to identify environmental carcinogens and gerontogens.[103] Cross-sectional studies are valuable for generating hypotheses, but longitudinal prospective studies are needed to establish strong associations and causality, and these studies would benefit from pairing

---

[103] Gerontogens are environmental agents that can accelerate aging.

humans with companion animals. Birnbaum suggested looking in both directions: adding pets to human studies and adding pet owners to longitudinal dog studies. She also noted the great potential of including sensors for personal monitoring.

These large studies will generate Big Data, and it is only going to get bigger and more complicated. Big cohort studies with tissue banks and data banks have real long-term benefits, said Birnbaum, who urged all players to work together: "We need to have everyone at the table," including public–private partnerships, and to use the entire arsenal of cross-disciplinary approaches. One Health should include public health, but it has been a challenge to involve the public health and medical communities. "It's bigger than we can address here, but we should be aware of it," she said. Untargeted analysis, suspect screening, and targeted analysis are all needed to identify environmental agents, she added. Practices and procedures in both data handling and biobanking need to be harmonized, and standardization of veterinary EMRs would help this effort.

To refine these studies, "we need to define what we mean by environment," said Birnbaum, noting that in this workshop, "environment" seemed to include all nongenetic factors while incorporating windows of susceptibility, effects of early life exposure, transgenerational effects, mixtures, interactions, and cumulative exposures. The genome, epigenome, microbiome, and metabolome act as both targets and mediators of environmental exposures. Considering the vastness of the exposome, Birnbaum wondered how necessary it was to analyze each exposure separately. "I think our companion animals are living in the same soup that we are living in, and rather than trying to break down what that soup is, maybe we need to start looking at the totality of the soup and how that affects the health of the dogs and cats and people," she suggested.

Our companion animals can inform us not only about cancer and aging but many other human health conditions, said Birnbaum, including those involving the immune system and the microbiome. Although several presentations emphasized the similarities of cancers in dogs and people, Birnbaum cautioned that neutered dogs may not be the best models for all diseases because the changed endocrine milieu may have a dramatic effect on the outcomes of some cancers.

Researchers also need to make more concerted efforts to reach out to disadvantaged communities. "We can't just build it and they will come," said Birnbaum, noting that dogs as well as people experience health disparities, and that greater diversification of participants will produce better-quality data. "The animal–human bond really defines our human society," added Birnbaum, but "dogs are not always household buddies," and working dogs may offer the opportunity to ask different questions. Although there is overlap between citizen science and CBPR, these are not the same thing. Academic or corporate researchers cannot helicopter into communities, get what they want, and leave. Developing a trusting relationship with communities can take time, which can be a problem for funders, but it is the reality of dealing with different communities, said Birnbaum. It is essential to communicate results back to the participants, even if the researchers do not know what the results mean. There also needs to be careful consideration of informed consent, privacy, and commercialization of data for both veterinary and human studies. "We have to be sure the work is not exploitative." Planning is key, and researchers should aim to involve an ethicist and an educator in the initial planning process, she added.

Reflecting on the presentations as a whole, Birnbaum emphasized that everyone comes to this from their own perspective. Clinicians, veterinarians, veterinary researchers, human health researchers, and exposure scientists all approached the work from a somewhat different angle. She reiterated a theme that had surfaced several times in the context of studying the health impacts of environmental exposures: "What is the question you are asking?"

Breen said, "The challenge is what to analyze, how to analyze, and how to merge the data. . . . It's the data management that's going to cause us a bit of a roadblock. Gathering the data is one thing, harnessing and using it is another." He observed that exposure research is very much bidirectional and he hoped that it will accelerate as a consequence of the workshop. "We do absolutely share exposures with our pets, and . . . the keys to unlocking some of the puzzles about environmental exposures have been walking alongside us for many years."

References

Angstadt, A. Y., V. Thayanithy, S. Subramanian, J. F. Modiano, and M. Breen. 2012. A genome-wide approach to comparative oncology: High-resolution oligonucleotide aCGH of canine and human osteosarcoma pinpoints shared microaberrations. *Cancer Genetics* 205(11):572–587.

Ashall, V., and P. Hobson-West. 2017. 'Doing good by proxy': Human-animal kinship and the 'donation' of canine blood. *Sociology of Health & Illness* 39(6):908–922.

Ashall, V., K. Millar, and P. Hobson-West. 2018. Informed consent in veterinary medicine: Ethical implications for the profession and the animal 'patient.' *Food Ethics* 1(3):247–258.

Aslan, B., L. Viola, S. S. Saini, J. Stockman, and E. P. Ryan. 2020. Pets as sentinels of human exposure to pesticides and co-exposure concerns with other contaminants/toxicants. In *Pets as sentinels, forecasters and promoters of human health*, edited by M. Pastorinho and A. Sousa. Cham, Switzerland: Springer.

Attfield, M. D., P. L. Schleiff, J. H. Lubin, A. Blair, P. A. Stewart, R. Vermeulen, J. B. Coble, and D. T. Silverman. 2012. The Diesel Exhaust in Miners Study: A cohort mortality study with emphasis on lung cancer. *Journal of the National Cancer Institute* 104(11):869–883.

Backer, L. C., A. M. Coss, A. F. Wolkin, W. D. Flanders, and J. S. Reif. 2008. Evaluation of associations between lifetime exposure to drinking water disinfection by-products and bladder cancer in dogs. *Journal of the American Veterinary Medical Association* 232(11):1663–1668.

Baldassarre, D., S. Castelnuovo, B. Frigerio, M. Amato, J. P. Werba, A. De Jong, A. L. Ravani, E. Tremoli, and C. R. Sirtori. 2009. Effects of timing and extent of smoking, type of cigarettes, and concomitant risk factors on the association between smoking and subclinical atherosclerosis. *Stroke* 40(6):1991–1998.

Barnes, D. E., and K. Yaffe. 2011. The projected effect of risk factor reduction on Alzheimer's disease prevalence. *The Lancet. Neurology* 10(9):819–828.

Becker, P. R., and S. A. Wise. 2006. The U.S. National Biomonitoring Specimen Bank and the Marine Environmental Specimen Bank. *Journal of Environmental Monitoring* 8(8):795–799. https://doi.org/10.1039/B602813F.

Beesoon, S., S. J. Genuis, J. P. Benskin, and J. W. Martin. 2012. Exceptionally high serum concentrations of perfluorohexanesulfonate in a Canadian family are linked to home carpet treatment applications. *Environmental Science & Technology* 46(23):12960–12967.

Belsky, D. W., A. Caspi, L. Arseneault, A. Baccarelli, D. L. Corcoran, X. Gao, E. Hannon, H. L. Harrington, L. J. Rasmussen, R. Houts, K. Huffman, W. E. Kraus, D. Kwon, J. Mill, C. F. Pieper, J. A. Prinz, R. Poulton, J. Schwartz, K. Sugden, P. Vokonas, B. S. Williams, and T. E. Moffitt. 2020. Quantification of the pace of biological aging in humans through a blood test, the DunedinPoAm DNA methylation algorithm. *eLife* 9:e54870.

Bernhardt, E. S., E. J. Rosi, and M. O. Gessner. 2017. Synthetic chemicals as agents of global change. *Frontiers in Ecology and the Environment* 15(2):84–90.

Bersimbaev, R., A. Pulliero, O. Bulgakova, K. Asia, A. Aripova, and A. Izzotti. 2020. Radon biomonitoring and microRNA in lung cancer. *International Journal of Molecular Sciences* 21(6):2154.

Bertone, E. R., L. A. Snyder, and A. S. Moore. 2002. Environmental tobacco smoke and risk of malignant lymphoma in pet cats. *American Journal of Epidemiology* 156(3):268–273.

Bertout, J. A., P. J. R. Baneux, and C. K. Robertson-Plouch. 2021. Recommendations for ethical review of veterinary clinical trials. *Frontiers in Veterinary Science* 8:715926.

Bishop, M. W., K. A. Janeway, and R. Gorlick. 2016. Future directions in the treatment of osteosarcoma. *Current Opinion in Pediatrics* 28(1):26–33.

Bost, P. C., M. J. Strynar, J. L. Reiner, J. A. Zweigenbaum, P. L. Secoura, A. B. Lindstrom, and J. A. Dye. 2016. U.S. domestic cats as sentinels for perfluoroalkyl substances: Possible linkages with housing, obesity, and disease. *Environmental Research* 151:145–153.

Bradley, R., I. Tagkopoulos, M. Kim, Y. Kokkinos, T. Panagiotakos, J. Kennedy, G. De Meyer, P. Watson, and J. Elliott. 2019. Predicting early risk of chronic kidney disease in cats using routine clinical laboratory tests and machine learning. *Journal of Veterinary Internal Medicine* 33(6):2644–2656.

Bujak, J. K., R. Pingwara, M. H. Nelson, and K. Majchrzak. 2018. Adoptive cell transfer: New perspective treatment in veterinary oncology. *Acta Veterinaria Scandinavica* 60(1):60.

Cacciottolo, M., X. Wang, I. Driscoll, N. Woodward, A. Saffari, J. Reyes, M. L. Serre, W. Vizuete, C. Sioutas, T. E. Morgan, M. Gatz, H. C. Chui, S. A. Shumaker, S. M. Resnick, M. A. Espeland, C. E. Finch, and J. C. Chen. 2017. Particulate air pollutants, APOE alleles and their contributions to cognitive impairment in older women and to amyloidogenesis in experimental models. *Translational Psychiatry* 7(1):e1022.

Calderón-Garciduñas, A. L., and C. Duyckaerts. 2017. Alzheimer disease. *Handbook of Clinical Neurology* 145:325–337.

Calderón-Garciduñas, L., A. Mora-Tiscareño, L. A. Fordham, C. J. Chung, R. García, N. Osnaya, J. Hernández, H. Acuña, T. M. Gambling, A. Villarreal-Calderón, J. Carson, H. S. Koren, and R. B. Devlin. 2001a. Canines as sentinel species for assessing chronic exposures to air pollutants: Part 1. Respiratory pathology. *Toxicological Sciences: An Official Journal of the Society of Toxicology* 61(2):342–355.

Calderón-Garciduñas, L., T. M. Gambling, H. Acuña, R. García, N. Osnaya, S. Monroy, A. Villarreal-Calderón, J. Carson, H. S. Koren, and R. B. Devlin. 2001b. Canines as sentinel species for assessing chronic exposures to air pollutants: Part 2. Cardiac pathology. *Toxicological Sciences: An Official Journal of the Society of Toxicology* 61(2):356–367.

Callier, V. 2019. Solving Peto's Paradox to better understand cancer. *Proceedings of the National Academy of Sciences* 116(6):1825-1828.

Campbell, L. D., J. J. Astrin, Y. DeSouza, J. Giri, A. A. Patel, M. Rawley-Payne, A. Rush, and N. Sieffert. 2018. The 2018 revision of the ISBER best practices: Summary of changes and the editorial team's development process. *Biopreservation and Biobanking* 16(1):3–6.

Catchpoole, D. 2016. 'Biohoarding': Treasures not seen, stories not told. *Journal of Health Services Research & Policy* 21(2):140–142.

Chambers, R. D., N. C. Yoder, A. B. Carson, C. Junge, D. E. Allen, L. M. Prescott, S. Bradley, G. Wymore, K. Lloyd, and S. Lyle. 2021. Deep learning classification of canine behavior using a single collar-mounted accelerometer: Real-world validation. *Animals: An Open Access Journal from MDPI* 11(6):1549.

Chen, B., C. Chen, and W. H. Hsu. 2015. Face recognition and retrieval using cross-age reference coding with cross-age celebrity dataset. *Institute of Electrical and Electronics Engineers (IEEE) Transactions on Multimedia* 17(6):804–815.

Chen, H., J. C. Kwong, R. Copes, K. Tu, P. J. Villeneuve, A. van Donkelaar, P. Hystad, R. V. Martin, B. J. Murray, B. Jessiman, A. S. Wilton, A. Kopp, and R. T. Burnett. 2017. Living near major roads and the incidence of dementia, Parkinson's disease, and multiple sclerosis: A population-based cohort study. *The Lancet* 389(10070):718–726.

Chung, M. K., K. Kannan, G. M. Louis, and C. J. Patel. 2018. Toward capturing the exposome: Exposure biomarker variability and coexposure patterns in the shared environment. *Environmental Science & Technology* 52(15):8801–8810.

Craun, K., J. Ekena, J. Sacco, T. Jiang, A. Motsinger-Reif, and L. A. Trepanier. 2020. Genetic and environmental risk for lymphoma in boxer dogs. *Journal of Veterinary Internal Medicine* 34(5):2068–2077.

Crimmins, E. M., J. K. Kim, and T. E. Seeman. 2009. Poverty and biological risk: The earlier "aging" of the poor. *The Journals of Gerontology. Series A, Biological Sciences and Medical Sciences* 64(2):286–292.

Culp, W. T. N., and R. Rebhun. 2020. Tumors of the respiratory system. In *Withrow and MacEwen's small animal clinical oncology*, 6th ed., edited by D. M. Vail, D. H. Thamm, and J. M. Liptak: Philadelphia, PA: W. B. Saunders. Pp. 507–514.

Deane-Coe, P. E., E. T. Chu, A. Slavney, A. R. Boyko, and A. J. Sams. 2018. Direct-to-consumer DNA testing of 6,000 dogs reveals 98.6-kb duplication associated with blue eyes and heterochromia in Siberian Huskies. *PLOS Genetics* 14(10):e1007648.

DeGregori, J. 2011. Evolved tumor suppression: Why are we so good at not getting cancer? *Cancer Research* 71(11):3739-3744.

Dietrich, H. G., and K. Golka. 2012. Bladder tumors and aromatic amines—Historical milestones from Ludwig Rehn to Wilhelm Hueper. *Frontiers in Bioscience* (*Elite edition*) 4(1):279–288.

Dignam, T., R. B. Kaufmann, L. Lestourgeon, and M. J. Brown. 2019. Control of lead sources in the United States, 1970–2017: Public health progress and current challenges to eliminating lead exposure. *Journal of Public Health Management & Practice* 25(Suppl 1):S13–S22. https://doi.org/10.1097/PHH.0000000000000889.

Dye, J. A., M. Venier, L. Zhu, C. R. Ward, R. A. Hites, and L. S. Birnbaum. 2007. Elevated PBDE levels in pet cats: Sentinels for humans? *Environmental Science & Technology* 41(18):6350–6356.

Dyer, J. L., and L. Milot. 2019. Social vulnerability assessment of dog intake location data as a planning tool for community health program development: A case study in Athens-Clarke County, GA, 2014–2016. *PLOS ONE* 14(12):e0225282.

Esposito, M., A. De Roma, P. Maglio, D. Sansone, G. Picazio, R. Bianco, C. De Martinis, G. Rosato, L. Baldi, and P. Gallo. 2019. Heavy metals in organs of stray dogs and cats from the city of Naples and its surroundings (Southern Italy). *Environmental Science and Pollution Research* 26(4):3473–3478. https://doi.org/10.1007/s11356-018-3838-5.

Eto, K., A. Yasutake, A. Nakano, H. Akagi, H. Tokunaga, and T. Kojima. 2001. Reappraisal of the historic 1959 cat experiment in Minamata by the Chisso factory. *The Tohoku Journal of Experimental Medicine* 194(4):197–203.

Evans, E. J. Jr., and J. DeGregori. 2021. Cells with cancer-associated mutations overtake our tissues as we age. *Aging and Cancer* 2(3):82–97.

Field, A. E., N. A. Robertson, T. Wang, A. Havas, T. Ideker, and P. D. Adams. 2018. DNA methylation clocks in aging: Categories, causes, and consequences. *Molecular Cell* 71(6):882–895.

Finch, C. E., and A. Haghani. 2021. Gene-environment interactions and stochastic variations in the gero-exposome. *The Journals of Gerontology. Series A, Biological Sciences and Medical Sciences* 76(10):1740–1747.

Fleming, T. P., A. J. Watkins, M. A. Velazquez, J. C. Mathers, A. M. Prentice, J. Stephenson, M. Barker, R. Saffery, C. S. Yajnik, J. J. Eckert, M. A. Hanson, T. Forrester, P. D. Gluckman, and K. M. Godfrey. 2018. Origins of lifetime health around the time of conception: Causes and consequences. *The Lancet* 391(10132):1842–1852.

Forman, H. J., and C. E. Finch. 2018. A critical review of assays for hazardous components of air pollution. *Free Radical Biology & Medicine* 117:202–217.

Forster, G. M., D. Hill, G. Gregory, K. M. Weishaar, S. Lana, J. E. Bauer, and E. P. Ryan. 2012. Effects of cooked navy bean powder on apparent total tract nutrient digestibility and safety in healthy adult dogs. *Journal of Animal Science* 90(8):2631–2638.

Forster, G. M., D. G. Brown, G. P. Dooley, R. L. Page, and E. P. Ryan. 2014. Multiresidue analysis of pesticides in urine of healthy adult companion dogs. *Environmental Science & Technology* 48(24):14677–14685.

Forster, G. M., A. L. Heuberger, C. D. Broeckling, J. E. Bauer, and E. P. Ryan. 2015. Consumption of cooked navy bean powders modulate the canine fecal and urine metabolome. *Current Metabolomics* 3(2):90–101.

Forster, G. M., J. Stockman, N. Noyes, A. L. Heuberger, C. D. Broeckling, C. M. Bantle, and E. P. Ryan. 2018. A comparative study of serum biochemistry, metabolome and microbiome parameters of clinically healthy, normal weight, overweight, and obese companion dogs. *Topics in Companion Animal Medicine* 33(4):126–135.

Fowler, B. L., C. M. Johannes, A. O'Connor, D. Collins, J. Lustgarten, C. Yuan, K. Weishaar, K. Sullivan, K. R. Hume, J. Mahoney, B. Vale, A. Schubert, V. Ball, K. Cooley-Lock, K. M. Curran, L. Nafe, A. Gedney, M. Weatherford, and D. N. LeVine. 2020. Ecological level analysis of primary lung tumors in dogs and cats and environmental radon activity. *Journal of Veterinary Internal Medicine* 34(6):2660–2670.

Freedman, L. P., I. M. Cockburn, and T. S. Simcoe. 2015. The economics of reproducibility in preclinical research. *PLOS Biology* 13(6):e1002165.

Furman, J. L., and S. Stern. 2011. Climbing atop the shoulders of giants: The impact of institutions on cumulative research. *American Economic Review* 101(5):1933–1963.

García-Guzmán, J. J., C. Pérez-Ràfols, M. Cuartero, and G. A. Crespo. 2021. Microneedle based electrochemical (bio)sensing: Towards decentralized and continuous health status monitoring. *Trends in Analytical Chemistry* 135(2021):116148.

Gatto, N. M., V. W. Henderson, H. N. Hodis, J. A. St. John, F. Lurmann, J. C. Chen, and W. J. Mack. 2014. Components of air pollution and cognitive function in middle-aged and older adults in Los Angeles. *Neurotoxicology* 40:1–7.

Gavazza, A., S. Presciuttini, R. Barale, G. Lubas, and B. Gugliucci. 2001. Association between canine malignant lymphoma, living in industrial areas, and use of chemicals by dog owners. *Journal of Veterinary Internal Medicine* 15(3):190–195.

Gavazza, A., G. Rossi, G. Lubas, M. Cerquetella, Y. Minamoto, and J. S. Suchodolski. 2018. Faecal microbiota in dogs with multicentric lymphoma. *Veterinary and Comparative Oncology* 16(1):E169–E175.

German, N. J., H. Yoon, R. Z. Yusuf, J. P. Murphy, L. W. S. Finley, G. Laurent, W. Haas, F. K. Satterstrom, J. Guarnerio, E. Zaganjor, D. Santos, P. P. Pandolfi, A. H. Beck, S. P. Gygi, D. T. Scadden, W. G. Kaelin, and M. C. Haigis. 2016. PHD3 loss in cancer enables metabolic reliance on fatty acid oxidation via deactivation of ACC2. *Molecular Cell* 63(6):1006–1020.

Giambò, F., M. Teodoro, C. Costa, and C. Fenga. 2021. Toxicology and microbiota: How do pesticides influence gut microbiota? A review. *International Journal of Environmental Research and Public Health* 18(11):5510.

Glickman, L. T., L. M. Domanski, T. G. Maguire, R. R. Dubielzig, and A. Churg. 1983. Mesothelioma in pet dogs associated with exposure of their owners to asbestos. *Environmental Research* 32(2):305–313.

Glickman, L. T., F. S. Schofer, L. J. McKee, J. S. Reif, and M. H. Goldschmidt. 1989. Epidemiologic study of insecticide exposures, obesity, and risk of bladder cancer in household dogs. *Journal of Toxicology and Environmental Health* 28(4):407–414.

Glickman, L. T., M. Raghavan, D. W. Knapp, P. L. Bonney, and M. H. Dawson. 2004. Herbicide exposure and the risk of transitional cell carcinoma of the urinary bladder in Scottish terriers. *Journal of the American Veterinary Medical Association* 224(8):1290–1297.

Goodson, W. H., 3rd, L. Lowe, D. O. Carpenter, M. Gilbertson, A. Manaf Ali, A. Lopez de Cerain Salsamendi, A. Lasfar, A. Carnero, A. Azqueta, A. Amedei, A. K. Charles, A. R. Collins, A. Ward, A. C. Salzberg, A. Colacci, A.-K. Olsen, A. Berg, B. J. Barclay, B. P. Zhou, C. Blanco-Aparicio, C. J. Baglole, C. Dong, C. Mondello, C.-W. Hsu, C. C. Naus, C. Yedjou, C. S. Curran, D. W. Laird, D. C. Koch, D. J. Carlin, D. W. Felsher, D. Roy, D. G. Brown, E. Ratovitski, E. P. Ryan, E. Corsini, E. Rojas, E.-Y. Moon, E. Laconi, F. Marongiu, F. Al-Mulla, F. Chiaradonna, F. Darroudi, F. L. Martin, F. J. Van Schooten, G. S. Goldberg, G. Wagemaker, G. N. Nangami, G. M. Calaf, G. Williams, G. T. Wolf, G. Koppen, G. Brunborg, H. K. Lyerly, H. Krishnan, H. Ab Hamid, H. Yasaei, H. Sone, H. Kondoh, H. K. Salem, H.-Y. Hsu, H. H. Park, I. Koturbash, I. R. Miousse, A. I. Scovassi, J. E. Klaunig, J. Vondráček, J. Raju, J. Roman, J. P. Wise, J. R. Whitfield, J. Woodrick, J. A. Christopher, J. Ochieng, J. F. Martinez-Leal, J. Weisz, J. Kravchenko, J. Sun, K. R. Prudhomme, K. B. Narayanan, K. A. Cohen-Solal, K. Moorwood, L. Gonzalez, L. Soucek, L. Jian, L. S. D'Abronzo, L.-T. Lin, L. Li, L. Gulliver, L. J. McCawley, L. Memeo, L. Vermeulen, L. Leyns, L. Zhang, M. Valverde, M. Khatami, M. F. Romano, M. Chapellier, M. A. Williams, M. Wade, M. H. Manjili, M. E. Lleonart, M. Xia, M. J. Gonzalez, M. V. Karamouzis, M. Kirsch-Volders, M. Vaccari, N. B. Kuemmerle, N. Singh, N. Cruickshanks, N. Kleinstreuer, N. van Larebeke, N. Ahmed, O. Ogunkua, P. K. Krishnakumar, P. Vadgama, P. A. Marignani, P. M. Ghosh, P. Ostrosky-Wegman, P. A. Thompson, P. Dent, P. Heneberg, P. Darbre, P. Sing Leung, P. Nangia-Makker, Q. S. Cheng, R. B. Robey, R. Al-Temaimi, R. Roy, R. Andrade-Vieira, R. K. Sinha, R. Mehta, R. Vento, R. Di Fiore, R. Ponce-Cusi, R. Dornetshuber-Fleiss, R. Nahta, R. C. Castellino, R. Palorini, R. Abd Hamid, S. A. S. Langie, S. E. Eltom, S. A. Brooks, S. Ryeom, S. S. Wise, S. N. Bay, S. A. Harris, S. Papagerakis, S. Romano, S. Pavanello, S. Eriksson, S. Forte, S. C. Casey, S. Luanpitpong, T.-J. Lee, T. Otsuki, T. Chen, T. Massfelder, T. Sanderson, T. Guarnieri, T. Hultman, V. Dormoy, V. Odero-Marah, V. Sabbisetti, V. Maguer-Satta, W. K. Rathmell, W. Engström, W. K. Decker, W. H. Bisson, Y. Rojanasakul, Y. Luqmani, Z. Chen, and Z. Hu. 2015. Assessing the carcinogenic potential of low-dose exposures to chemical mixtures in the environment: The challenge ahead. *Carcinogenesis* 36:S254–S296.

Gray, C., M. Fox, and P. Hobson-West. 2018. Reconciling autonomy and beneficence in treatment decision-making for companion animal patients. *The Liverpool Law Review* 39(1):47–69.

Grieco, V., E. Riccardi, G. F. Greppi, F. Teruzzi, V. Iermanò, and M. Finazzi. 2008a. Canine testicular tumours: A study on 232 dogs. *Journal of Comparative Pathology* 138(2–3):86–89.

Grieco, V., E. Riccardi, M. C. Veronesi, C. Giudice, and M. Finazzi. 2008b. Evidence of testicular dysgenesis syndrome in the dog. *Theriogenology* 70(1):53–60.

Gu, D., Z. L. Neuman, J. F. Modiano, and R. J. Turesky. 2012. Biomonitoring the cooked meat carcinogen 2-amino-1-methyl-6-phenylimidazo[4,5-b]pyridine in canine fur. *Journal of Agricultural and Food Chemistry* 60(36):9371–9375.

Guo, W., J.-S. Park, Y. Wang, S. Gardner, C. Baek, M. Petreas, and K. Hooper. 2012. High polybrominated diphenyl ether levels in California house cats: House dust a primary source? *Environmental Toxicology and Chemistry* 31(2):301–306.

Guo, W., S. Gardner, S. Yen, M. Petreas, and J.-S. Park. 2016. Temporal changes of PBDE levels in California house cats and a link to cat hyperthyroidism. *Environmental Science & Technology* 50(3):1510–1518.

Gustafson, D. L., D. L. Duval, D. P. Regan, and D. H. Thamm. 2018. Canine sarcomas as a surrogate for the human disease. *Pharmacology & Therapeutics* 188:80–96.

Guy, M. K., R. L. Page, W. A. Jensen, P. N. Olson, J. D. Haworth, E. E. Searfoss, and D. E. Brown. 2015. The Golden Retriever Lifetime Study: Establishing an observational cohort study with translational relevance for human health. *Philosophical Transactions of the Royal Society of London. Series B, Biological Sciences* 370(1673):20140230.

Haghani, A., M. Cacciottolo, K. R. Doty, C. D'Agostino, M. Thorwald, N. Safi, M. E. Levine, C. Sioutas, T. C. Town, H. J. Forman, H. Zhang, T. E. Morgan, and C. E. Finch. 2020. Mouse brain transcriptome responses to inhaled nanoparticulate matter differed by sex and APOE in Nrf2-Nfkb interactions. *eLife* 9:e54822.

Hammel, S. C., K. Hoffman, T. F. Webster, K. A. Anderson, and H. M. Stapleton. 2016. Measuring personal exposure to organophosphate flame retardants using silicone wristbands and hand wipes. *Environmental Science & Technology* 50(8):4483–4491.

Harbison, M. L., and J. J. Godleski. 1983. Malignant mesothelioma in urban dogs. *Veterinary Pathology* 20(5):531–540.

Hayes, H. M., R. Hoover, and R. E. Tarone. 1981. Bladder cancer in pet dogs: A sentinel for environmental cancer? *American Journal of Epidemiology* 114(2):229–233.

Hayes, H. M., R. E. Tarone, K. P. Cantor, C. R. Jessen, D. M. McCurnin, and R. C. Richardson. 1991. Case-control study of canine malignant lymphoma: Positive association with dog owner's use of 2,4-dichlorophenoxyacetic acid herbicides. *Journal of the National Cancer Institute* 83(17):1226–1231.

Hayward, J. J., M. G. Castelhano, K. C. Oliveira, E. Corey, C. Balkman, T. L. Baxter, M. L. Casal, S. A. Center, M. Fang, S. J. Garrison, S. E. Kalla, P. Korniliev, M. I. Kotlikoff, N. S. Moise, L. M. Shannon, K. W. Simpson, N. B. Sutter, R. J. Todhunter, and A. R. Boyko. 2016. Complex disease and phenotype mapping in the domestic dog. *Nature Communications* 7:10460.

Henry, C. J., M. Casás-Selves, J. Kim, V. Zaberezhnyy, L. Aghili, A. E. Daniel, L. Jimenez, T. Azam, E. N. McNamee, E. T. Clambey, J. Klawitter, N. J. Serkova, A. C. Tan, C. A. Dinarello, and J. DeGregori. 2015. Aging-associated inflammation promotes selection for adaptive oncogenic events in B cell progenitors. *The Journal of Clinical Investigation* 125(12):4666–4680.

Hewitt, R., and P. Watson. 2013. Defining biobank. *Biopreservation and Biobanking* 11(5):309–315.

Hoffman, K., J. L. Levasseur, S. Zhang, D. Hay, N. J. Herkert, and H. M. Stapleton. 2021. Monitoring human exposure to organophosphate esters: Comparing silicone wristbands with spot urine samples as predictors of internal dose. *Environmental Science & Technology Letters* 8(9):805–810.

Hofmann, J. N., L. E. Beane Freeman, K. Murata, G. Andreotti, J. J. Shearer, K. Thoren, L. Ramanathan, C. G. Parks, S. Koutros, C. C. Lerro, D. Liu, N. Rothman, C. F. Lynch, B. I. Graubard, D. P. Sandler, M. C. Alavanja, and O. Landgren. 2021. Lifetime pesticide use and monoclonal gammopathy of undetermined significance in a prospective cohort of male farmers. *Environmental Health Perspectives* 129(1):17003.

Horvath, S., K. Singh, K. Raj, S. Khairnar, A. Sanghavi, A. Shrivastava, J. A. Zoller, C. Z. Li, C. Herenu, M. Canatelli-Mallat, M. Lehmann, L. C. Solberg Woods, A. Garcia Martinez, T. Wang, P. Chiavellini, A. J. Levine, H. Chen, R. G. Goya, and H. L. Katcher. 2020. Reversing age: Dual species measurement of epigenetic age with a single clock (preprint, not certified by peer review). *bioRxiv* 2020.05.07.082917.

Islami, F., E. M. Ward, H. Sung, K. A. Cronin, F. Tangka, R. L. Sherman, J. Zhao, R. N. Anderson, S. J. Henley, K. R. Yabroff, A. Jemal, and V. B. Benard. 2021. Annual report to the nation on the status of cancer, Part 1: National cancer statistics. *Journal of the National Cancer Institute* 113(12):1648–1669.

James, A. K., S. Nehzati, N. V. Dolgova, D. Sokaras, T. Kroll, K. Eto, J. L. O'Donoghue, G. E. Watson, G. J. Myers, P. H. Krone, I. J. Pickering, and G. N. George. 2020. Rethinking the Minamata tragedy: What mercury species was really responsible? *Environmental Science & Technology* 54(5):2726–2733.

Jones, H., and K. Rowland-Yeo. 2013. Basic concepts in physiologically based pharmacokinetic modeling in drug discovery and development. *CPT: Pharmacometrics & Systems Pharmacology* 2(8):e63.

Jones, R. R., G. Hoek, J. A. Fisher, S. Hasheminassab, D. Wang, M. H. Ward, C. Sioutas, R. Vermeulen, and D. T. Silverman. 2020. Land use regression models for ultrafine particles, fine particles, and black carbon in Southern California. *The Science of the Total Environment* 699:134234.

Kalia, V., M. M. Niedzwiecki, J. M. Bradner, F. K. Lau, M. L. Bucher, K. E. Manz, Z. Coates Fuentes, K. D. Pennell, M. Picard, D. I. Walker, W. T. Hu, D. P. Jones, and G. W. Miller. 2021. Cross-species metabolomic analysis of DDT and Alzheimer's disease-associated tau toxicity (preprint, not certified by peer review). *bioRxiv* 2021.06.14.448355.

Karthikraj, R., R. Bollapragada, and K. Kannan. 2018a. Melamine and its derivatives in dog and cat urine: An exposure assessment study. *Environmental Pollution* 238:248–254.

Karthikraj, R., S. Borkar, S. Lee, and K. Kannan. 2018b. Parabens and their metabolites in pet food and urine from New York State, United States. *Environmental Science & Technology* 52(6):3727–3737.

Karthikraj, R., and K. Kannan. 2019. Widespread occurrence of glyphosate in urine from pet dogs and cats in New York State, USA. *The Science of the Total Environment* 659:790–795.

Karthikraj, R., S. Lee, and K. Kannan. 2019. Urinary concentrations and distribution profiles of 21 phthalate metabolites in pet cats and dogs. *The Science of the Total Environment* 690:70–75.

Karyadi, D. M., E. Karlins, B. Decker, B. M. vonHoldt, G. Carpintero-Ramirez, H. G. Parker, R. K. Wayne, and E. A. Ostrander. 2013. A copy number variant at the KITLG locus likely confers risk for canine squamous cell carcinoma of the digit. *PLOS Genetics* 9(3):e1003409.

Katz, T. A., S. L. Grimm, A. Kaushal, J. Dong, L. S. Treviño, R. K. Jangid, A. V. Gaitán, J.-P. Bertocchio, Y. Guan, M. J. Robertson, R. M. Cabrera, M. J. Finegold, C. E. Foulds, C. Coarfa, and C. L. Walker. 2020. Hepatic tumor formation in adult mice developmentally exposed to organotin. *Environmental Health Perspectives* 128(1):17010.

Kawakami, T., M. K. Jensen, A. Slavney, P. E. Deane, A. Milano, V. Raghavan, B. Ford, E. T. Chu, A. J. Sams, and A. R. Boyko. 2021. R-locus for roaned coat is associated with a tandem duplication in an intronic region of USH2A in dogs and also contributes to Dalmatian spotting. *PLOS ONE* 16(3):e0248233.

Kerr, K. R., G. Forster, S. E. Dowd, E. P. Ryan, and K. S. Swanson. 2013. Effects of dietary cooked navy bean on the fecal microbiome of healthy companion dogs. *PLOS ONE* 8(9):e74998.

Kim, H. T., J. P. Loftus, S. Mann, and J. J. Wakshlag. 2018. Evaluation of arsenic, cadmium, lead and mercury contamination in over-the-counter available dry dog foods with different animal ingredients (red meat, poultry, and fish). *Frontiers in Veterinary Science* 5:264.

Kirkness, E. F., V. Bafna, A. L. Halpern, S. Levy, K. Remington, D. B. Rusch, A. L. Delcher, M. Pop, W. Wang, C. M. Fraser, and J. C. Venter. 2003. The dog genome: Survey sequencing and comparative analysis. *Science* 301(5641):1898–1903.

Knize, M. G., C. P. Salmon, and J. S. Felton. 2003. Mutagenic activity and heterocyclic amine carcinogens in commercial pet foods. *Mutation Research* 539(1–2):195–201.

Kochmanski, J., L. Montrose, J. M. Goodrich, and D. C. Dolinoy. 2017. Environmental deflection: The impact of toxicant exposures on the aging epigenome. *Toxicological Sciences: An Official Journal of the Society of Toxicology* 156(2):325–335.

Künzli, N., M. Jerrett, R. Garcia-Esteban, X. Basagaña, B. Beckermann, F. Gilliland, M. Medina, J. Peters, H. N. Hodis, and W. J. Mack. 2010. Ambient air pollution and the progression of atherosclerosis in adults. *PLOS ONE* 5(2):e9096.

Langlois, D. K., J. B. Kaneene, V. Yuzbasiyan-Gurkan, B. L. Daniels, H. Mejia-Abreu, N. A. Frank, and J. P. Buchweitz. 2017. Investigation of blood lead concentrations in dogs living in Flint, Michigan. *Journal of the American Veterinary Medical Association* 251(8):912–921.

Lauby-Secretan, B., C. Scoccianti, D. Loomis, Y. Grosse, F. Bianchini, and K. Straif, for the International Agency for Research on Cancer Handbook Working Group. 2016. Body fatness and cancer—viewpoint of the IARC Working Group. *The New England Journal of Medicine* 375(8):794–798. https://dx.doi.org/10.1056%2FNEJMsr1606602.

Lea, R. G., A. S. Byers, R. N. Sumner, S. M. Rhind, Z. Zhang, S. L. Freeman, R. Moxon, H. M. Richardson, M. Green, J. Craigon, and G. C. W. England. 2016. Environmental chemicals impact dog semen quality in vitro and may be associated with a temporal decline in sperm motility and increased cryptorchidism. *Scientific Reports* 6:31281. Erratum in: *Scientific Reports* 6:33267.

LeBlanc, A. K., and C. N. Mazcko. 2020. Improving human cancer therapy through the evaluation of pet dogs. *Nature Reviews Cancer* 20(12):727–742.

LeBlanc, A. K., C. N. Mazcko, A. Cherukuri, E. P. Berger, W. C. Kisseberth, M. E. Brown, S. E. Lana, K. Weishaar, B. K. Flesner, J. N. Bryan, D. M. Vail, J. H. Burton, J. L. Willcox, A. J. Mutsaers, J. P. Woods, N. C. Northrup, C. Saba, K. M. Curran, H. Leeper, H. Wilson-Robles, B. G. Wustefeld-Janssens, S. Lindley, A. N. Smith, N. Dervisis, S. Klahn, M. L. Higginbotham, R. M. Wouda, E. Krick, J. A. Mahoney, C. A.

London, L. G. Barber, C. E. Balkman, A. L. McCleary-Wheeler, S. E. Suter, O. Martin, A. Borgatti, K. Burgess, M. O. Childress, J. L. Fidel, S. D. Allstadt, D. L. Gustafson, L. E. Selmic, C. Khanna, and T. M. Fan. 2021. Adjuvant sirolimus does not improve outcome in pet dogs receiving standard-of-care therapy for appendicular osteosarcoma: A prospective, randomized trial of 324 dogs. *Clinical Cancer Research: An Official Journal of the American Association for Cancer Research* 27(11):3005–3016.

Le Moal, J., S. Goria, A. Guillet, A. Rigou, and J. Chesneau. 2021. Time and spatial trends of operated cryptorchidism in France and environmental hypotheses: A nationwide study from 2002 to 2014. *Human Reproduction* 36(5):1383–1394.

Lermen, D., F. Gwinner, M. Bartel-Steinbach, S. C. Mueller, J. K. Habermann, M.-B. Balwir, E. Smits, A. Virgolino, U. Fiddicke, M. Berglund, A. Åkesson, A. Bergstrom, K. Leander, M. Horvat, J. Snoj Tratnik, M. Posada de la Paz, A. Castaño Calvo, M. Esteban López, H. von Briesen, H. Zimmermann, and M. Kolossa-Gehring. 2020. Towards harmonized biobanking for biomonitoring: A comparison of human biomonitoring-related and clinical biorepositories. *Biopreservation and Biobanking* 18(2):122–135.

Levine, H., N. Jørgensen, A. Martino-Andrade, J. Mendiola, D. Weksler-Derri, I. Mindlis, R. Pinotti, and S. H. Swan. 2017. Temporal trends in sperm count: A systematic review and meta-regression analysis. *Human Reproduction Update* 23(6):646–659.

Lindblad-Toh, K., C. M. Wade, T. S. Mikkelsen, E. K. Karlsson, D. B. Jaffe, M. Kamal, M. Clamp, J. L. Chang, E. J. Kulbokas, M. C. Zody, E. Mauceli, X. Xie, M. Breen, R. K. Wayne, E. A. Ostrander, C. P. Ponting, F. Galibert, D. R. Smith, P. J. DeJong, E. Kirkness, P. Alvarez, T. Biagi, W. Brockman, J. Butler, C.-W. Chin, A. Cook, J. Cuff, M. J. Daly, D. DeCaprio, S. Gnerre, M. Grabherr, M. Kellis, M. Kleber, C. Bardeleben, L. Goodstadt, A. Heger, C. Hitte, L. Kim, K.-P. Koepfli, H. G. Parker, J. P. Pollinger, S. M. J. Searle, N. B. Sutter, R. Thomas, C. Webber, J. Baldwin, A. Abebe, A. Abouelleil, L. Aftuck, M. Ait-Zahra, T. Aldredge, N. Allen, P. An, S. Anderson, C. Antoine, H. Arachchi, A. Aslam, L. Ayotte, P. Bachantsang, A. Barry, T. Bayul, M. Benamara, A. Berlin, D. Bessette, B. Blitshteyn, T. Bloom, J. Blye, L. Boguslavskiy, C. Bonnet, B. Boukhgalter, A. Brown, P. Cahill, N. Calixte, J. Camarata, Y. Cheshatsang, J. Chu, M. Citroen, A. Collymore, P. Cooke, T. Dawoe, R. Daza, K. Decktor, S. DeGray, N. Dhargay, K. Dooley, K. Dooley, P. Dorje, K. Dorjee, L. Dorris, N. Duffey, A. Dupes, O. Egbiremolen, R. Elong, J. Falk, A. Farina, S. Faro, D. Ferguson, P. Ferreira, S. Fisher, M. FitzGerald, K. Foley, C. Foley, A. Franke, D. Friedrich, D. Gage, M. Garber, G. Gearin, G. Giannoukos, T. Goode, A. Goyette, J. Graham, E. Grandbois, K. Gyaltsen, N. Hafez, D. Hagopian, B. Hagos, J. Hall, C. Healy, R. Hegarty, T. Honan, A. Horn, N. Houde, L. Hughes, L. Hunnicutt, M. Husby, B. Jester, C. Jones, A. Kamat, B. Kanga, C. Kells, D. Khazanovich, A. C. Kieu, P. Kisner, M. Kumar, K. Lance, T. Landers, M. Lara, W. Lee, J.-P. Leger, N. Lennon, L. Leuper, S. LeVine, J. Liu, X. Liu, Y. Lokyitsang, T. Lokyitsang, A. Lui, J. Macdonald, J. Major, R. Marabella, K. Maru, C. Matthews, S. McDonough, T. Mehta, J. Meldrim, A. Melnikov, L. Meneus, A. Mihalev, T. Mihova, K. Miller, R. Mittelman, V. Mlenga, L. Mulrain, G. Munson, A. Navidi, J. Naylor, T. Nguyen, N. Nguyen, C. Nguyen, T. Nguyen, R. Nicol, N. Norbu, C. Norbu, N. Novod, T. Nyima, P. Olandt, B. O'Neill, K. O'Neill, S. Osman, L. Oyono, C. Patti, D. Perrin, P. Phunkhang, F. Pierre, M. Priest, A. Rachupka, S. Raghuraman, R. Rameau, V. Ray, C. Raymond, F. Rege, C. Rise, J. Rogers, P. Rogov, J. Sahalie, S. Settipalli, T. Sharpe, T. Shea, M. Sheehan, N. Sherpa, J. Shi, D. Shih,

J. Sloan, C. Smith, T. Sparrow, J. Stalker, N. Stange-Thomann, S. Stavropoulos, C. Stone, S. Stone, S. Sykes, P. Tchuinga, P. Tenzing, S. Tesfaye, D. Thoulutsang, Y. Thoulutsang, K. Topham, I. Topping, T. Tsamla, H. Vassiliev, V. Venkataraman, A. Vo, T. Wangchuk, T. Wangdi, M. Weiand, J. Wilkinson, A. Wilson, S. Yadav, S. Yang, X. Yang, G. Young, Q. Yu, J. Zainoun, L. Zembek, A. Zimmer, and E. S. Lander. 2005. Genome sequence, comparative analysis and haplotype structure of the domestic dog. *Nature* 438(7069):803–819.

Liu, K. H., C. M. Lee, G. Singer, P. Bais, F. Castellanos, M. H. Woodworth, T. R. Ziegler, C. S. Kraft, G. W. Miller, S. Li, Y. M. Go, E. T. Morgan, and D. P. Jones. 2021. Large scale enzyme based xenobiotic identification for exposomics. *Nature Communications* 12(1):5418.

Ly, L. H., E. Gordon, and A. Protopopova. 2021. Exploring the relationship between human social deprivation and animal surrender to shelters in British Columbia, Canada. *Frontiers in Veterinary Science* 8:656597.

Ma, J., H. Zhu, and K. Kannan. 2020. Fecal excretion of perfluoroalkyl and polyfluoroalkyl substances in pets from New York State, United States. *Environmental Science & Technology Letters* 7(3):135–142.

Macías-Montes, A., M. Zumbado, O. P. Luzardo, Á. Rodríguez-Hernández, A. Acosta-Dacal, C. Rial-Berriel, L. D. Boada, and L. A. Henríquez-Hernández. 2021. Nutritional evaluation and risk assessment of the exposure to essential and toxic elements in dogs and cats through the consumption of pelleted dry food: How important is the quality of the feed? *Toxics* 9(6):133.

Martin, K. M., M. A. Rossing, L. M. Ryland, R. F. DiGiacomo, and W. A. Freitag. 2000. Evaluation of dietary and environmental risk factors for hyperthyroidism in cats. *Journal of the American Veterinary Medical Association* 217(6):853–856.

Martincorena, I., K. M. Raine, M. Gerstung, K. J. Dawson, K. Haase, P. Van Loo, H. Davies, M. R. Stratton, and P. J. Campbell. 2017. Universal patterns of selection in cancer and somatic tissues. *Cell* 171(5):1029–1041.e21. Erratum in: *Cell* 173(7):1823.

Matchock, R. L. 2015. Pet ownership and physical health. *Current Opinion in Psychiatry* 28(5):386–392.

Mattingly, C. J., T. E. McKone, M. A. Callahan, J. A. Blake, and E. A. C. Hubal. 2012. Providing the missing link: The exposure science ontology ExO. *Environmental Science & Technology* 46(6):3046–3053.

Mattson, M. P., V. D. Longo, and M. Harvie. 2017. Impact of intermittent fasting on health and disease processes. *Ageing Research Reviews* 39:46–58.

McHale, C. M., G. Osborne, R. Morello-Frosch, A. G. Salmon, M. S. Sandy, G. Solomon, L. Zhang, M. T. Smith, and L. Zeise. 2018. Assessing health risks from multiple environmental stressors: Moving from GxE to IxE. *Mutation Research. Reviews in Mutation Research* 775:11–20.

Messier, K. P., D. C. Wheeler, A. R. Flory, R. R. Jones, D. Patel, B. T. Nolan, and M. H. Ward. 2019. Modeling groundwater nitrate exposure in private wells of North Carolina for the Agricultural Health Study. *The Science of the Total Environment* 655:512–519.

Middleton, R. P., S. Lacroix, M.-P. Scott-Boyer, N. Dordevic, A. D. Kennedy, A. R. Slusky, J. Carayol, C. Petzinger-Germain, A. Beloshapka, and J. Kaput. 2017. Metabolic differences between dogs of different body sizes. *Journal of Nutrition and Metabolism* 2017:4535710.

Miller, G. W., and D. P. Jones. 2014. The nature of nurture: Refining the definition of the exposome. *Toxicological Sciences: An Official Journal of the Society of Toxicology* 137(1):1–2.

Mochizuki, H., K. Kennedy, S. G. Shapiro, and M. Breen. 2015. BRAF mutations in canine cancers. *PLOS ONE* 10(6):e0129534.

Monsalve, S., J. Hammerschmidt, M. L. Izar, S. Marconcin, F. Rizzato, G. Polo, and R. Garcia. 2018. Associated factors of companion animal neglect in the family environment in Pinhais, Brazil. *Preventive Veterinary Medicine* 157:19–25.

Montrose, L., C. W. Noonan, Y. H. Cho, J. Lee, J. Harley, T. O'Hara, C. Cahill, and T. J. Ward. 2015. Evaluating the effect of ambient particulate pollution on DNA methylation in Alaskan sled dogs: Potential applications for a sentinel model of human health. *The Science of the Total Environment* 512–513:489–494.

Mor, D. E., S. Sohrabi, R. Kaletsky, W. Keyes, A. Tartici, V. Kalia, G. W. Miller, and C. T. Murphy. 2020. Metformin rescues Parkinson's disease phenotypes caused by hyperactive mitochondria. *Proceedings of the National Academy of Sciences of the United States of America* 117(42):26438–26447.

Mousseau, T. A., and A. P. Møller. 2013. Elevated frequency of cataracts in birds from Chernobyl. *PLOS ONE* 8(7):e66939.

Mouttham, L., S. J. Garrison, D. L. Archer, and M. G. Castelhano. 2021. A biobank's journey: Implementation of a quality management system and accreditation to ISO 20387. *Biopreservation and Biobanking* 19(3):163–170.

NASEM (National Academies of Sciences, Engineering, and Medicine). 2015. *The role of clinical studies for pets with naturally occurring tumors in translational cancer research*: Workshop summary. Washington, DC: The National Academies Press.

NASEM. 2018. *Learning through citizen science: Enhancing opportunities by design.* Washington, DC: The National Academies Press.

NCI (National Cancer Institute). 2016. *NCI Best Practices for Biospecimen Resources*, 3rd ed. https://biospecimens.cancer.gov/bestpractices/2016-NCIBestPractices.pdf (accessed February 28, 2022).

NRC (National Research Council). 1991. *Animals as sentinels of environmental health hazards.* Washington, DC: The National Academies Press.

O'Connell, S. G., L. D. Kincl, and K. A. Anderson. 2014. Silicone wristbands as personal passive samplers. *Environmental Science & Technology* 48(6):3327–3335.

Otto, C. M., E. Hare, J. P. Buchweitz, K. M. Kelsey, and S. D. Fitzgerald. 2020. Fifteen-year surveillance of pathological findings associated with death or euthanasia in search-and-rescue dogs deployed to the September 11, 2001, terrorist attack sites. *Journal of the American Veterinary Medical Association* 257(7):734–743.

Paoloni, M., and C. Khanna. 2008. Translation of new cancer treatments from pet dogs to humans. *Nature Reviews Cancer* 8(2):147–156.

Pastor, M., K. Chalvet-Monfray, T. Marchal, G. Keck, J. P. Magnol, C. Fournel-Fleury, and F. Ponce. 2009. Genetic and environmental risk indicators in canine non-Hodgkin's lymphomas: Breed associations and geographic distribution of 608 cases diagnosed throughout France over 1 year. *Journal of Veterinary Internal Medicine* 23(2):301–310.

Paynter, A. N., M. D. Dunbar, K. E. Creevy, and A. Ruple. 2021. Veterinary big data: When data goes to the dogs. *Animals: An Open Access Journal from MDPI* 11(7):1872.

Perera, B. P. U., C. Faulk, L. K. Svoboda, J. M. Goodrich, and D. C. Dolinoy. 2020. The role of environmental exposures and the epigenome in health and disease. *Environmental and Molecular Mutagenesis* 61(1):176–192.

Peterson, M. E. 2014. Animal models of disease: Feline hyperthyroidism: An animal model for toxic nodular goiter. *The Journal of Endocrinology* 223(2):T97–T114.

Peto, R. 2015. Quantitative implications of the approximate irrelevance of mammalian body size and lifespan to lifelong cancer risk. *Philosophical Transactions of the Royal Society B: Biological Sciences* 370(1673):20150198.

Peto, R. 2016. Epidemiology, multistage models, and short-term mutagenicity tests. *International Journal of Epidemiology* 45(3):621-637.

Pontius, J. U., J. C. Mullikin, D. R. Smith, Agencourt Sequencing Team, K. Lindblad-Toh, S. Gnerre, M. Clamp, J. Chang, R. Stephens, B. Neelam, N. Volfovsky, A. A. Schäffer, R. Agarwala, K. Narfström, W. J. Murphy, U. Giger, A. L. Roca, A. Antunes, M. Menotti-Raymond, N. Yuhki, J. Pecon-Slattery, W. E. Johnson, G. Bourque, G. Tesler, N. C. S. Program, and S. J. O'Brien. 2007. Initial sequence and comparative analysis of the cat genome. *Genome Research* 17(11):1675–1689.

Poutasse, C. M., J. B. Herbstman, M. E. Peterson, J. Gordon, P. H. Soboroff, D. Holmes, D. Gonzalez, L. G. Tidwell, and K. A. Anderson. 2019. Silicone pet tags associate tris(1,3-dichloro-2-isopropyl) phosphate exposures with feline hyperthyroidism. *Environmental Science & Technology* 53(15):9203–9213.

Raghavan, M., D. W. Knapp, P. L. Bonney, M. H. Dawson, and L. T. Glickman. 2005. Evaluation of the effect of dietary vegetable consumption on reducing risk of transitional cell carcinoma of the urinary bladder in Scottish Terriers. *Journal of the American Veterinary Medical Association* 227(1):94–100.

Rappaport, S. M., D. K. Barupal, D. Wishart, P. Vineis, and A. Scalbert. 2014. The blood exposome and its role in discovering causes of disease. *Environmental Health Perspectives* 122:769–774.

Reese, S. W., K. H. Tully, J. Nabi, M. Paciotti, W. H. Chou, and Q.-D. Trinh. 2021. Temporal trends in the incidence of testicular cancer in the United States over the past four decades. *European Urology Oncology* 4(5):834–836.

Reif, J. S., K. Dunn, G. K. Ogilvie, and C. K. Harris. 1992. Passive smoking and canine lung cancer risk. *American Journal of Epidemiology* 135(3):234–239.

Reif, J. S., C. Bruns, and K. S. Lower. 1998. Cancer of the nasal cavity and paranasal sinuses and exposure to environmental tobacco smoke in pet dogs. *American Journal of Epidemiology* 147(5):488–492.

Rial-Berriel, C., L.A. Henríquez-Hernández, and O.P. Luzardo. 2020. Role of pet dogs and cats as sentinels of human exposure to polycyclic aromatic hydrocarbons. In *Pets as sentinels, forecasters and promoters of human health*, edited by M. Pastorinho and A. Sousa: Cham, Switzerland: Springer.

Ringel, A. E., J. M. Drijvers, G. J. Baker, A. Catozzi, J. C. García-Cañaveras, B. M. Gassaway, B. C. Miller, V. R. Juneja, T. H. Nguyen, S. Joshi, C.-H. Yao, H. Yoon, P. T. Sage, M. W. LaFleur, J. D. Trombley, C. A. Jacobson, Z. Maliga, S. P. Gygi, P. K. Sorger, J. D. Rabinowitz, A. H. Sharpe, and M. C. Haigis. 2020. Obesity shapes metabolism in the tumor microenvironment to suppress anti-tumor immunity. *Cell* 183(7):1848–1866.e26.

Rozhok, A., and J. DeGregori. 2019. A generalized theory of age-dependent carcinogenesis. *eLife* 8:e39950.
Ruple, A., A. C. Avery, and P. S. Morley. 2017. Differences in the geographic distribution of lymphoma subtypes in golden retrievers in the USA. *Veterinary and Comparative Oncology* 15(4):1590–1597.
Ruple, A., B. N. Bonnett, and R. L. Page. 2019. Epidemiology and the evidence-based medicine approach. In *Withrow and MacEwen's small animal clinical oncology*, 6th ed., edited by D. M. Vail, D. H. Thamm, and J. M. Liptak: Philadelphia, PA: W. B. Saunders. Pp. 81–97.
Samant, P. P., M. M. Niedzwiecki, N. Raviele, V. Tran, J. Mena-Lapaix, D. I. Walker, E. I. Felner, D. P. Jones, G. W. Miller, and M. R. Prausnitz. 2020. Sampling interstitial fluid from human skin using a microneedle patch. *Science Translational Medicine* 12(571):eaaw0285.
Samet, J. M., E. Avila-Tang, P. Boffetta, L. M. Hannan, S. Olivo-Marston, M. J. Thun, and C. M. Rudin. 2009. Lung cancer in never smokers: Clinical epidemiology and environmental risk factors. *Clinical Cancer Research: An Official Journal of the American Association for Cancer Research* 15(18):5626–5645.
Schecter, A., O. Päpke, K. C. Tung, J. Joseph, T. R. Harris, and J. Dahlgren. 2005. Polybrominated diphenyl ether flame retardants in the U.S. population: Current levels, temporal trends, and comparison with dioxins, dibenzofurans, and polychlorinated biphenyls. *Journal of Occupational and Environmental Medicine* 47(3):199–211.
Schiffman, J. D., and M. Breen. 2015. Comparative oncology: What dogs and other species can teach us about humans with cancer. *Philosophical Transactions of the Royal Society of London. Series B, Biological Sciences* 370(1673):20140231.
Schmitz, L. L., W. Zhao, S. M. Ratliff, J. Goodwin, J. Miao, Q. Lu, X. Guo, K. D. Taylor, J. Ding, Y. Liu, M. Levine, and J. A. Smith. 2021. The socioeconomic gradient in epigenetic ageing clocks: Evidence from the Multi-Ethnic Study of Atherosclerosis and the Health and Retirement Study. *Epigenetics* Jul 6:1–23.
Selmic, L. E., J. H. Burton, D. H. Thamm, S. J. Withrow, and S. E. Lana. 2014. Comparison of carboplatin and doxorubicin-based chemotherapy protocols in 470 dogs after amputation for treatment of appendicular osteosarcoma. *Journal of Veterinary Internal Medicine* 28(2):554–563.
Shannon, L. M., R. H. Boyko, M. Castelhano, E. Corey, J. J. Hayward, C. McLean, M. E. White, M. A. Said, B. A. Anita, N. I. Bondjengo, J. Calero, A. Galov, M. Hedimbi, B. Imam, R. Khalap, D. Lally, A. Masta, K. C. Oliveira, L. Pérez, J. Randall, N. M. Tam, F. J. Trujillo-Cornejo, C. Valeriano, N. B. Sutter, R. J. Todhunter, C. D. Bustamante, and A. R. Boyko. 2015. Genetic structure in village dogs reveals a central Asian domestication origin. *Proceedings of the National Academy of Sciences* 112(44):13639-13644.
Shefchek, K. A., N. L. Harris, M. Gargano, N. Matentzoglu, D. Unni, M. Brush, D. Keith, T. Conlin, N. Vasilevsky, X. A. Zhang, J. P. Balhoff, L. Babb, S. M. Bello, H. Blau, Y. Bradford, S. Carbon, L. Carmody, L. E. Chan, V. Cipriani, A. Cuzick, M. Della Rocca, N. Dunn, S. Essaid, P. Fey, C. Grove, J. P. Gourdine, A. Hamosh, M. Harris, I. Helbig, M. Hoatlin, M. Joachimiak, S. Jupp, K. B. Lett, S. E. Lewis, C. McNamara, Z. M. Pendlington, C. Pilgrim, T. Putman, V. Ravanmehr, J. Reese, E. Riggs, S. Robb, P. Roncaglia, J. Seager, E. Segerdell, M. Similuk, A. L. Storm, C. Thaxon, A. Thessen,

J. O. B. Jacobsen, J. A. McMurry, T. Groza, S. Köhler, D. Smedley, P. N. Robinson, C. J. Mungall, M. A. Haendel, M. C. Munoz-Torres, and D. Osumi-Sutherland. 2020. The Monarch Initiative in 2019: An integrative data and analytic platform connecting phenotypes to genotypes across species. *Nucleic Acids Research* 48(D1):D704–D715.

Simpson, S., M. D. Dunning, S. de Brot, L. Grau-Roma, N. P. Mongan, and C. S. Rutland. 2017. Comparative review of human and canine osteosarcoma: Morphology, epidemiology, prognosis, treatment and genetics. *Acta Veterinaria Scandinavica* 59(1):71.

Skinner, C. G., J. D. Thomas, and J. D. Osterloh. 2010. Melamine toxicity. *Journal of Medical Toxicology: Official Journal of the American College of Medical Toxicology* 6(1):50–55.

Smedley, D., M. Schubach, J. O. B. Jacobsen, S. Köhler, T. Zemojtel, M. Spielmann, M. Jäger, H. Hochheiser, N. L. Washington, J. A. McMurry, M. A. Haendel, C. J. Mungall, S. E. Lewis, T. Groza, G. Valentini, and P. N. Robinson. 2016. A whole-genome analysis framework for effective identification of pathogenic regulatory variants in Mendelian disease. *American Journal of Human Genetics* 99(3):595–606.

Smith, M. T., K. Z. Guyton, C. F. Gibbons, J. M. Fritz, C. J. Portier, I. Rusyn, D. M. DeMarini, J. C. Caldwell, R. J. Kavlock, P. F. Lambert, S. S. Hecht, J. R. Bucher, B. W. Stewart, R. A. Baan, V. J. Cogliano, and K. Straif. 2016. Key characteristics of carcinogens as a basis for organizing data on mechanisms of carcinogenesis. *Environmental Health Perspectives* 124(6):713–721.

Smith, N., K. R. Luethcke, K. Craun, and L. Trepanier. 2021. Risk of bladder cancer and lymphoma in dogs is associated with pollution indices by county of residence. *Veterinary and Comparative Oncology* 20(1):246–255.

Soleri, D., J. Long, M. D. Ramirez-Andreotta, R. Eitemiller, and R. Pandya. 2016. Finding pathways to more equitable and meaningful public-scientist partnerships. *Citizen Science: Theory and Practice* 1(1):9.

Spinelli, J. B., H. Yoon, A. E. Ringel, S. Jeanfavre, C. B. Clish, and M. C. Haigis. 2017. Metabolic recycling of ammonia via glutamate dehydrogenase supports breast cancer biomass. *Science* 358(6365):941–946.

Stern, Y., C. Habeck, J. Steffener, D. Barulli, Y. Gazes, Q. Razlighi, D. Shaked, and T. Salthouse. 2014. The Reference Ability Neural Network Study: Motivation, design, and initial feasibility analyses. *NeuroImage* 103:139–151.

Stoffel, E. M., and C. C. Murphy. 2020. Epidemiology and mechanisms of the increasing incidence of colon and rectal cancers in young adults. *Gastroenterology* 158(2):341–353.

Strynar, M. J., and A. B. Lindstrom. 2008. Perfluorinated compounds in house dust from Ohio and North Carolina, USA. *Environmental Science & Technology* 42(10):3751–3756.

Sumner, R. N., M. Tomlinson, J. Craigon, G. C. W. England, and R. G. Lea. 2019. Independent and combined effects of diethylhexyl phthalate and polychlorinated biphenyl 153 on sperm quality in the human and dog. *Scientific Reports* 9(1):3409.

Sumner, R. N., I. T. Harris, M. Van der Mescht, A. Byers, G. C. W. England, and R. G. Lea. 2020. The dog as a sentinel species for environmental effects on human fertility. *Reproduction* 159(6):R265–R276.

Sung, J., S. Kim, J. J. T. Cabatbat, S. Jang, Y.-S. Jin, G. Y. Jung, N. Chia, and P.-J. Kim. 2017. Global metabolic interaction network of the human gut microbiota for context-specific community-scale analysis. *Nature Communications* 8:15393.

Takashima-Uebelhoer, B. B., L. G. Barber, S. E. Zagarins, E. Procter-Gray, A. L. Gollenberg, A. S. Moore, and E. R. Bertone-Johnson. 2012. Household chemical exposures and the risk of canine malignant lymphoma, a model for human non-Hodgkin's lymphoma. *Environmental Research* 112:171–176.

Tang, L., G. R. Zirpoli, K. Guru, K. B. Moysich, Y. Zhang, C. B. Ambrosone, and S. E. McCann. 2010. Intake of cruciferous vegetables modifies bladder cancer survival. *Cancer Epidemiology, Biomarkers & Prevention: A Publication of the American Association for Cancer Research, Cosponsored by the American Society of Preventive Oncology* 19(7):1806–1811.

Tawa, G. J., J. Braisted, D. Gerhold, G. Grewal, C. Mazcko, M. Breen, G. Sittampalam, and A. K. LeBlanc. 2021. Transcriptomic profiling in canines and humans reveals cancer specific gene modules and biological mechanisms common to both species. *PLOS Computational Biology* 17(9):e1009450.

Thessen, A. E., R. L. Walls, L. Vogt, J. Singer, R. Warren, P. L. Buttigieg, J. P. Balhoff, C. J. Mungall, D. L. McGuinness, B. J. Stucky, M. J. Yoder, and M. A. Haendel. 2020. Transforming the study of organisms: Phenomic data models and knowledge bases. *PLOS Computational Biology* 16(11):e1008376.

Tierney, L. A., F. F. Hahn, and J. F. Lechner. 1996. p53, erbB-2 and K-ras gene alterations are rare in spontaneous and plutonium-239-induced canine lung neoplasia. *Radiation Research* 145(2):181–187.

Tindle, H. A., M. Stevenson Duncan, R. A. Greevy, R. S. Vasan, S. Kundu, P. P. Massion, and M. S. Freiberg. 2018. Lifetime smoking history and risk of lung cancer: Results from the Framingham Heart Study. *Journal of the National Cancer Institute* 110(11):1201–1207.

Trinh, P., J. R. Zaneveld, S. Safranek, and P. M. Rabinowitz. 2018. One Health relationships between human, animal, and environmental microbiomes: A mini-review. *Frontiers in Public Health* 6:235.

Trumble, B. C., and C. E. Finch. 2019. The exposome in human evolution: From dust to diesel. *The Quarterly Review of Biology* 94(4):333–394.

Tsai, J., D. M. Homa, A. S. Gentzke, M. Mahoney, S. R. Sharapova, C. S. Sosnoff, K. T. Caron, L. Wang, P. C. Melstrom, and K. F. Trivers. 2018. Exposure to secondhand smoke among nonsmokers—United States, 1988–2014. *MMWR: Morbidity and Mortality Weekly Report* 67:1342–1346. http://dx.doi.org/10.15585/mmwr.mm6748a3.

Urfer, S. R., M. Darvas, K. Czeibert, S. Sándor, D. E. L. Promislow, K. E. Creevy, E. Kubinyi, and M. Kaeberlein. 2021. Canine cognitive dysfunction (CCD) scores correlate with amyloid beta 42 levels in dog brain tissue. *GeroScience* 43(5):2379–2386.

Vardarajan, B., V. Kalia, J. Manly, A. Brickman, D. Reyes-Dumeyer, R. Lantigua, I. Ionita-Laza, D. P. Jones, G. W. Miller, and R. Mayeux. 2020. Differences in plasma metabolites related to Alzheimer's disease, APOE ε4 status, and ethnicity. *Alzheimer's & Dementia* 6(1):e12025.

Verhoeven, J. E., R. Yang, O. M. Wolkowitz, F. S. Bersani, D. Lindqvist, S. H. Mellon, R. Yehuda, J. D. Flory, J. Lin, D. Abu-Amara, I. Makotkine, C. Marmar, M. Jett, and R. Hammamieh. 2018. Epigenetic age in male combat-exposed war veterans: Associations with posttraumatic stress disorder status. *Molecular Neuropsychiatry* 4(2):90–99.

Vermeulen, R., E. L. Schymanski, A.-L. Barabási, and G. W. Miller. 2020. The exposome and health: Where chemistry meets biology. *Science* 367(6476):392–396.

von Hippel, F. A., E. J. Trammell, J. Merilä, M. B. Sanders, T. Schwarz, J. H. Postlethwait, T. A. Titus, C. L. Buck, and I. Katsiadaki. 2016. The ninespine stickleback as a model organism in Arctic ecotoxicology. *Evolutionary Ecology Research* 17:487–504.

Wakabayashi, K., M. Nagao, H. Esumi, and T. Sugimura. 1992. Food-derived mutagens and carcinogens. *Cancer Research* 52(7 Suppl):2092s–2098s.

Walker, C. L., and S.-m. Ho. 2012. Developmental reprogramming of cancer susceptibility. *Nature Reviews Cancer* 12(7):479–486.

Walker, D. I., B. D. Juran, A. C. Cheung, E. M. Schlicht, Y. Liang, M. Niedzwiecki, N. F. LaRusso, G. J. Gores, D. P. Jones, G. W. Miller, and K. N. Lazaridis. 2021. High-resolution exposomics and metabolomics reveals specific associations in cholestatic liver diseases. *Hepatology Communications* Nov 26.

Walker, R. L., and J. A. Fisher. 2018. Companion animal studies: Slipping through a research oversight gap. *The American Journal of Bioethics: AJOB* 18(10):62–63.

Wallerstein, N., and B. Duran. 2010. Community-based participatory research contributions to intervention research: The intersection of science and practice to improve health equity. *American Journal of Public Health* 100(Suppl 1):S40–S46.

Wang, T., J. Ma, A. N. Hogan, S. Fong, K. Licon, B. Tsui, J. F. Kreisberg, P. D. Adams, A.-R. Carvunis, D. L. Bannasch, E. A. Ostrander, and T. Ideker. 2020. Quantitative translation of dog-to-human aging by conserved remodeling of the DNA methylome. *Cell Systems* 11(2):176–185.e6.

Wang, Z., R. Altenburger, T. Backhaus, A. Covaci, M. L. Diamond, J. O. Grimalt, R. Lohmann, A. Schäffer, M. Scheringer, H. Selin, A. Soehl, and N. Suzuki. 2021. We need a global science-policy body on chemicals and waste. *Science* 371(6531):774–776.

Ward, Z. J., S. N. Bleich, A. L. Cradock, J. L. Barrett, C. M. Giles, C. Flax, M. W. Long, and S. L. Gortmaker. 2019. Projected U.S. state-level prevalence of adult obesity and severe obesity. *The New England Journal of Medicine* 381(25):2440–2450.

Watowich, M. M., E. L. MacLean, B. Hare, J. Call, J. Kaminski, Á. Miklósi, and N. Snyder-Mackler. 2020. Age influences domestic dog cognitive performance independent of average breed lifespan. *Animal Cognition* 23(4):795–805.

WHO (World Health Organization). 2018. *Burden of disease from household air pollution for 2016*. https://www.who.int/airpollution/data/HAP_BoD_results_May2018_final.pdf (accessed February 28, 2022).

Wild, C. P. 2005. Complementing the genome with an "exposome": The outstanding challenge of environmental exposure measurement in molecular epidemiology. *Cancer Epidemiology, Biomarkers & Prevention: A Publication of the American Association for Cancer Research, Cosponsored by the American Society of Preventive Oncology* 14(8):1847–1850.

Wise, C. F., S. C. Hammel, N. Herkert, J. Ma, A. Motsinger-Reif, H. M. Stapleton, and M. Breen. 2020. Comparative exposure assessment using silicone passive samplers indicates that domestic dogs are sentinels to support human health research. *Environmental Science & Technology* 54(12):7409–7419.

Wise, C. F., S. C. Hammel, N. J. Herkert, M. Ospina, A. M. Calafat, M. Breen, and H. M. Stapleton. 2022. Comparative assessment of pesticide exposures in domestic dogs and their owners using silicone passive samplers and biomonitoring. *Environmental Science & Technology* 56(2):1149–1161.

Witkop, J. J., T. Vertigan, A. Reynolds, L. Duffy, B. Barati, S. Jerome, and K. Dunlap. 2021. Sled dogs as a model for PM2.5 exposure from wildfires in Alaska. *Environment International* 156:106767.

Worgul, B. V., Y. I. Kundiyev, N. M. Sergiyenko, V. V. Chumak, P. M. Vitte, C. Medvedovsky, E. V. Bakhanova, A. K. Junk, O. Y. Kyrychenko, N. V. Musijachenko, S. A. Shylo, O. P. Vitte, S. Xu, X. Xue, and R. E. Shore. 2007. Cataracts among Chernobyl cleanup workers: Implications regarding permissible eye exposures. *Radiation Research* 167(2):233-243.

Xing, J.-S., and Z.-M. Bai. 2018. Is testicular dysgenesis syndrome a genetic, endocrine, or environmental disease, or an unexplained reproductive disorder? *Life Sciences* 194:120–129.

Yeates, J., and J. Savulescu. 2017. Companion animal ethics: A special area of moral theory and practice? *Ethical Theory and Moral Practice* 20(2):347–359.

Yordy, J., C. Kraus, J. J. Hayward, M. E. White, L. M. Shannon, K. E. Creevy, D. E. L. Promislow, and A. R. Boyko. 2020. Body size, inbreeding, and lifespan in domestic dogs. *Conservation Genetics* 21(1):137–148.

Zhan, J., Y. Liang, D. Liu, X. Ma, P. Li, C. Liu, X. Liu, P. Wang, and Z. Zhou. 2018. Antibiotics may increase triazine herbicide exposure risk via disturbing gut microbiota. *Microbiome* 6(1):224.

Zhang, J., L. Wang, and K. Kannan. 2019. Polyethylene terephthalate and polycarbonate microplastics in pet food and feces from the United States. *Environmental Science & Technology* 53(20):12035–12042.

Appendix A

Statement of Task

A planning committee of the National Academies of Sciences, Engineering, and Medicine will organize and host a 1.5-day public workshop that will examine the potential role of companion animals as sentinels of relevant, shared environmental exposures that may affect human aging and cancer. The workshop will explore the opportunities and challenges for using this novel translational approach to exposure science as a way to accelerate the knowledge turn in this evolving field. The workshop will feature invited presentations and panel discussions on topics that may include:

- Potential data sources needed to assess whether companion animals may serve as sentinels for human environmental exposures.
- The state of the science for biomarkers of exposure and use of biosensors for application to companion animal populations of interest.
- Best practices for collection, storage, and analysis of biosamples to assess exposures (e.g., biorepository resources, DNA susceptibility, DNA methylation, microbiome, etc.).
- Strategies for standardizing, sharing, and aggregating health records and relevant metadata across species.
- Current policies and regulations related to monitoring and mitigating environmental exposures and the role for prospective interventions based on companion animal data.

The planning committee will develop the agenda for the workshop sessions, select and invite speakers and discussants, and moderate the discussions. A proceedings of the presentations and discussions at the workshop will be prepared by a designated rapporteur in accordance with institutional guidelines.

Appendix B

Workshop Agenda

DECEMBER 1, 2021

7:30 am   **Registration and Breakfast**
Outside Keck 100 & E-Street Conference Room

8:00 am   **Welcome and Workshop Overview**
Linda Birnbaum, Planning Committee Chair and Emeritus Director, National Institute of Environmental Health Sciences (NIEHS) and National Toxicology Program (NTP)

Ned Sharpless, Director, National Cancer Institute (NCI)

8:30 am   **Session 1: History and Current State of the Science of Environmental Exposure Effects on Aging and Cancer Susceptibility**
Co-Moderators: Bill Farland, Colorado State University, Emeritus & Danielle Carlin, NIEHS*

State of the Science and Overview
- **Environmental Exposure and Cancer**
Gary Ellison, NIEHS/NCI

---

*=presented remotely

- **Environmental Exposure and Cancer in Companion Animals**
  Audrey Ruple, Virginia Tech

- **Aging Targets for Environmental Exposure**
  Marcia Haigis, Harvard University
  Cheryl Walker, Baylor College of Medicine*

- **Canine Genetic Systems and Relevance for Human Cancers**
  Elaine Ostrander, National Human Genome Research Institute

10:15 am Break

10:30 am Session 1: History and Current State of the Science of Environmental Exposure Effects on Aging and Cancer Susceptibility
Co-Moderators: Bill Farland, Colorado State University, Emeritus & Danielle Carlin, NIEHS*

Aging and Cancer Susceptibility
James DeGregori, University of Colorado Cancer Center

What Are the Gaps in Human Cancer Prevention and Control That May Be Addressed Through Companion Animal Research, and Vice Versa?
Peter Rabinowitz, University of Washington*

Session 1 Panel Discussion

12:15 pm Lunch Break with Virtual Poster Session
E-Street Conference Room & Zoom

1:30 pm Session 2a: Methods and Current Studies
Moderator: Myrtle Davis, Bristol Myers Squibb

- **Exposome and Health**
  Gary Miller, Columbia University

- **Biomonitoring of Chemical Exposure in Companion Animals**
  Kurunthachalam Kannan, New York University

- **Assessing the Exposome Using Wearable Sensors: Challenges and Opportunities**
  *Yuxia Cui, NIEHS**

- **Ongoing Canine Population Studies**
  Matthew Breen, North Carolina State University

3:00 pm   Break

3:20 pm   Session 2a (cont'd): Methods and Current Studies
Moderator: Myrtle Davis, Bristol Myers Squibb

Ongoing Canine Population Studies (cont'd)
Rod Page, Colorado State University
Daniel Promislow, University of Washington
Richard Lea, University of Nottingham

Session 2a   Panel Discussion

5:30 pm   Adjourn Day 1

Evening Reception with In-Person Poster Session
E-Street Conference Room & Balcony

DECEMBER 2, 2021

7:30 am   Breakfast
E-Street Conference Room

8:00 am   Session 2b: Relevance of Companion Animal Exposures to Human Cancer and Aging
Moderator: Nicole Deziel, Yale School of Public Health

- **Overview of Outdoor Air, Water, Ground**
  Caleb "Tuck" Finch, University of Southern California

- **A Comparative Assessment of SVOC Exposures in Domestic Dogs and Their Owners Using Silicone Passive Samplers**
  Heather Stapleton, Duke University

- **Indoor Products: Endocrine Disruptors, Flame Retardants, PFAS**
  Jan Dye, Environmental Protection Agency*

- **Radon Exposures**
  Chad Johannes, Iowa State University

- **Heavy Metal Exposures**
  Norman Kleiman, Columbia University

9:45 am Break

10:00 am Session 2b (cont'd): Relevance of Companion Animal Exposures to Human Cancer and Aging
Moderator: Nicole Deziel, Yale School of Public Health

- **Dietary and Feeding Exposures**
  Joe Wakshlag, Cornell University

- **Pesticides/Herbicides and Mixtures**
  Elizabeth Ryan, Colorado State University

Session 2b   Panel Discussion

12:00 pm Lunch Break with Virtual Poster Session
E-Street Conference Room & Zoom

1:15 pm Session 3: Accelerating Cross-species Comparisons: Opportunities and Challenges in Data Sources, Collection, Storage, Modeling, and Sharing
Moderator: Roy Jensen, University of Kansas*

- **Human Exposure Assessment**
  Rena Jones, NCI

- **Data and Sample Collection/Storage and Sharing/Data Integration of Human and Companion Animal Data**
  Amy K. LeBlanc, NCI
  Marta Castelhano, Cornell University
  Anne Thessen, University of Colorado Anschutz Medical Campus

2:40 pm	**Break**
3:00 pm	**Session 3 (cont'd): Accelerating Cross-Species Comparisons: Opportunities and Challenges in Data Sources, Collection, Storage, Modeling, and Sharing** *Moderator: Roy Jensen, University of Kansas**

- **Data, Samples, and Modeling**
  Angela Hughes, Mars Petcare
  Adam Boyko, Embark Veterinary, Inc./Cornell University
  Mark Dunn, American Kennel Club

Session 3   Panel Discussion

5:00 pm	**Adjourn Day 2**

Evening Reception with In-Person Poster Session
E-Street Conference Room & Balcony

DECEMBER 3, 2021

8:30 am	**Breakfast** E-Street Conference Room
9:00 am	**Session 4: Equity, Ethics, and Policy** *Moderator: Wendy Shelton, Virtual Beast/Colorado State University*

- **Ethical Considerations: Research Subject Protections, Citizen Science Issues, and Shared Health**
  Lisa Moses, Harvard Medical School*

- **One Health Approaches in Arctic Indigenous Communities**
  Frank A. von Hippel, University of Arizona

- **Aligning Health Care for a Bonded Family Society**
  Michael Blackwell, University of Tennessee, Knoxville*

Session 4   Panel Discussion

10:45 am **Break and Pick Up Box Lunches**
E Street Conference Room

11:00 am **Facilitated Discussions with Planning Committee Members**
E Street Conference Room

Objectives:
Review key messages from the workshop discussions, including identifying potential next steps, promising areas for future action, and opportunities for collaboration.
- *TABLE 1: Bill Farland, Colorado State University, Emeritus*
- *TABLE 2: Nicole Deziel, Yale School of Public Health*
- *TABLE 3: Wendy Shelton, Virtual Beast/Colorado State University*
- *TABLE 4: Daniel Promislow, University of Washington*
- *TABLE 5: Matthew Breen, North Carolina State University*
- *TABLE 6: Rod Page, Colorado State University*
- *VIRTUAL TABLE: Danielle Carlin, NIEHS, and Roy Jensen, University of Kansas**

11:45 am **Session 5: Identifying Research Gaps and Setting a Research Agenda: Recommendations and Next Steps for the Path Forward**
Moderator: Linda Birnbaum, NIEHS, NTP, Emeritus

Report Backs from Facilitated Discussions and Workshop Reflections (each with 5 minutes for report back summaries and workshop reflections, followed by open discussion among workshop participants)
- *Bill Farland, Colorado State University, Emeritus*
- *Nicole Deziel, Yale School of Public Health*
- *Wendy Shelton, Virtual Beast/Colorado State University*
- *Daniel Promislow, University of Washington*
- *Matthew Breen, North Carolina State University*
- *Rod Page, Colorado State University*
- *Danielle Carlin, NIEHS**
- *Roy Jensen, University of Kansas**

Concluding Remarks and Next Steps for the Field
Linda Birnbaum, NIEHS, NTP, Emeritus

1:00 pm **Adjourn**

Appendix C

Biographical Sketches of Planning Committee Members and Workshop Speakers

Linda S. Birnbaum, Ph.D., DABT, ATS, is the former director of the National Institute of Environmental Health Sciences (NIEHS) of the National Institutes of Health, and the National Toxicology Program (NTP). After retirement, she was granted scientist emeritus status and still maintains a laboratory. As a board-certified toxicologist, Dr. Birnbaum served as a federal scientist for 40 years. Prior to her appointment as NIEHS and NTP director in 2009, she spent 19 years at the U.S. Environmental Protection Agency (EPA), where she directed the largest division focusing on environmental health research. Dr. Birnbaum has received many awards and recognitions. In 2016, she was awarded the North Carolina Award in Science. She was elected to the Institute of Medicine of the National Academies, one of the highest honors in the fields of medicine and health. She was also elected to the Collegium Ramazzini, an independent, international academy comprised of internationally renowned experts in the fields of occupational and environmental health and received an honorary doctor of science from the University of Rochester and a Distinguished Alumna Award from the University of Illinois. She has also received honorary doctorates from the University of Rhode Island; Ben-Gurion University, Israel; and Amity University, India; the Surgeon General's Medallion 2014; and 14 Scientific and Technological Achievement Awards, which reflect the recommendations of EPA's external Science Advisory Board, for specific publications. She has also received numerous awards from professional societies and citizens' groups. Dr. Birnbaum is an active member of the scientific community. She was vice president of the International Union of Toxicology, the umbrella organization for toxicology societies in more than 50 countries, and former president of the Society of Toxicology, the largest professional organization of toxicologists in the world. She is the author of more than 1,000 peer-reviewed publications, book chapters, abstracts, and reports. Dr. Birnbaum's own research focuses on the pharmacokinetic behavior of environmental chemicals, mechanisms of action of toxicants including endocrine disruption, and linking of real-world exposures to health effects. She is an adjunct professor at the University of Queensland in Australia, the School of Public Health of Yale University, the Gillings School of Global Public Health, the Curriculum in Toxicology, and the Department of Environmental Sciences and Engineering at the University of North Carolina at Chapel Hill, as well as in the Integrated Toxicology and Environmental Health Program at Duke University where she is also a Scholar in Residence. A native of New Jersey, Dr. Birnbaum received her M.S. and Ph.D. in microbiology from the University of Illinois at Urbana-Champaign.

Michael J. Blackwell, D.V.M., M.P.H., FNAP, currently serves as the director of the Program for Pet Health Equity at the University of Tennessee. His mission is to improve access to veterinary care, especially for families with limited means. Previous to this position, Dr. Blackwell served as: dean, College of Veterinary Medicine, University of Tennessee; chief of staff, Office of the Surgeon General of the United States; deputy director, Center for Veterinary Medicine, Food and Drug Administration; and chief veterinary officer, U.S. Public Health Service. During 23 years on active duty with the U.S. Public Health Service, he achieved the rank of assistant surgeon general/rear admiral. Dr. Blackwell has received numerous awards and recognitions, including the Distinguished Service Medal, Meritorious Service Medal, and two Surgeon General's Exemplary Service Medals. He is the 2020 recipient of the Avanzino Leadership Award and the 2021 Senator John Melcher, DVM Leadership in Public Policy awardee.

Adam Boyko, Ph.D., M.S., is an associate professor in biomedical sciences at the Cornell University College of Veterinary Medicine, conducting research on canine genetics. He is also cofounder and chief science officer of Embark Veterinary, a dog DNA testing company founded in 2015 and incubated at the Cornell McGovern Center, and a trustee for the Morris Animal Foundation. His research focuses on complex trait mapping, bioinformatics, statistical genetics, inference of evolutionary forces and demographic history from genomic data, and understanding the evolutionary process of domestication and rapid adaptation. Prior to joining the faculty at Cornell, Dr. Boyko received undergraduate degrees in computer science and evolutionary ecology from the University of Illinois as well as a master's in computer science and a doctorate in biology at Purdue. He also worked as a postdoc and research associate at Cornell and Stanford, studying computational biology and population genomics.

Matthew Breen, Ph.D., C. Biol, FRSB, is a professor of genomics and the Oscar J. Fletcher Distinguished Professor of Comparative Oncology Genetics in the Department of Molecular Biomedical Sciences at the North Carolina State University (NCSU) College of Veterinary Medicine. He is also a member of the NCSU Comparative Medicine Institute (CMI), the Center for Human Health and the Environment, and the Genetics and Genomics Initiative, as well as the Duke Cancer Institute, and the Cancer Genetics Program at the University of North Carolina's Lineberger Comprehensive Cancer Center. Dr. Breen is a member of the NCSU Research Leadership Academy. Dr. Breen's research focuses on genetics, genomics, and the comparative aspects of animal and human health. The lab uses a range of genetic and genomic technologies for evaluating changes to genome structure that occur in canine cancers. With these data the lab aims to improve outcomes for canine cancer patients and also advance our understanding of the comparable cancers in people. In addition, the lab is assessing the impact of environmental exposures on animal health, as a sentinel for human health. He was a charter member, and serves on the board of directors, of the Canine Comparative Oncology and Genomics Consortium (CCOGC), a 501(c)(3) not-for-profit organization established to promote the role of the dog in comparative biomedical research, and also serves on the board of directors of the Canines-N-Kids Foundation, a 501(c)(3) committed to finding a cure to the devastating cancers that canines and children face and have in common. He is member of the steering committee of the National Cancer Institute's Integrated Canine Data Commons and serves on the Data Governance Advisory Board of that initiative. Dr. Breen was appointed to the National Academies expert committee tasked with planning a public workshop to examine the role of companion animals as sentinels of shared environmental exposures that may impact human aging and cancer. Dr. Breen has served on scientific review committees for organizations including the National Institutes of Health, the AKC Canine Health Foundation, and the Morris Animal Foundation. He is a regular reviewer for numerous scientific funding agencies and journals and serves on the editorial board of several journals.

Danielle Carlin, Ph.D., DABT, is a program administrator with the Superfund Research Program (SRP). Her position consists of providing guidance and advice to grantees applying for SRP P42 Center grants, and serving as the lead liaison between SRP trainees and the various training opportunities offered by SRP. She also oversees the xenobiotic metabolism and asbestos grant portfolios (e.g., R01s). Her current research interests include chemical mixtures, combined exposures, metals, asbestos, and xenobiotic metabolism.

Prior to her current position, she was a postdoctoral researcher for 4 years at the University of North Carolina: 2 years within the Eshelman School of Pharmacy, Division of Molecular Pharmaceutics, studying aerosolized drugs/vaccines for treatment and prevention of tuberculosis and 2 years within the Curriculum in Toxicology, conducting her research at the U.S. Environmental Protection Agency, in Research Triangle Park, NC, where she studied the toxicological effects of exposure to Libby amphibole asbestos in the rat model. Her areas of expertise include cardiopulmonary/reproductive physiology and inhalation toxicology/pharmacology. She received her Ph.D. in 2005 from Kansas State University, College of Veterinary Medicine, Department of Anatomy and Physiology. She also has a B.S. and an M.S. in animal science from New Mexico State University.

Marta Castelhano, D.V.M., MVSc, received her doctor of veterinary medicine and master of veterinary science degrees from the University of Lisbon, Portugal. Serving as an associate research professor at Cornell University, the director of the Cornell Veterinary Biobank (CVB), and at the Dog Aging Project Biobank, she has over 15 years of experience in the standardized collection, processing, storage, and distribution of high-quality biospecimens and associated data. Dr. Castelhano is a member of the Education and Training Committee at the International Society for Biological and Environmental Repositories (ISBER), where she creates educational opportunities for biobankers worldwide, and has contributed to the writing of the fourth edition of the *ISBER Best Practices: Recommendations for Repositories*. A frequent speaker at biobank conferences and symposiums, Dr. Castelhano was invited by the National Institutes of Standards and Technology (NIST) to represent the U.S. position in biobanking as an ISO expert and delegate. With her contribution, *ISO 20387: General Requirements for Biobanking* was the first ISO standard created specifically for biobanks. In April 2019, Dr. Castelhano led the CVB through third-party conformity assessment by the American Association of Laboratory Accreditation to become the first biobank in the world to receive accreditation to the ISO 20387 standard. She also serves in the ISBER COVID-19 task force, assessing the needs of biobankers worldwide during the pandemic, to inform the next generation of standard documents and to create targeted improvement opportunities, particularly for biobanks with limited resources.

Yuxia Cui, Ph.D., is a health scientist administrator at the National Institute of Environmental Health Sciences (NIEHS). Dr. Cui oversees the exposure science and the exposome grant portfolio that is focused on emerging technologies toward improved exposure and risk assessment in environmental health research. These include sensor technologies, '-omics-based approaches, computational and informatics-based methodologies, as well as other innovative approaches to enable an integrated view and better understanding of the exposome. Dr. Cui currently serves as the program officer for the laboratory network of the Human Health Exposure Analysis Resource. She is also a member of the National Institutes of Health Common Fund Metabolomics Program leadership team and oversees the day-to-day operations of the program. Dr. Cui received training in molecular toxicology and transcriptomics and received her doctorate in environmental toxicology from Duke University.

Myrtle A. Davis, D.V.M., Ph.D., is the vice president of discovery toxicology at Bristol Myers Squibb (BMS). Myrtle joined BMS from the National Cancer Institute, where she was the chief of the Toxicology and Pharmacology Branch of the Developmental Therapeutics Program. Dr. Davis has previous experience as a research advisor in the Drug Safety group of Lilly Research Laboratories. In both roles, she contributed critical expertise to the advancement of several drugs candidates and to the understanding of toxicological mechanisms. She also has several years of academic experience as an associate professor in the Department of Pathology in the School of Medicine at the University of Maryland. Dr. Davis is currently responsible for leading the scientific efforts in discovery toxicology to provide target and molecular hazard identification and risk assessments for issues identified in discovery research. She also leads and oversees the investigative toxicology efforts needed to support mechanistic understanding of compound- or target-mediated toxicities in discovery and development. Dr. Davis is a Fellow of the Academy of Toxicological Sciences, an active member of the Society of Toxicology (recently elected as vice president–elect for the Society), and a member of the Society of Toxicologic Pathology. She is currently serving on the Board of Scientific Councilors of the National Toxicology Program, and she is a reviewer for the Assay Development and Screening Technologies Laboratory of the National Center for Advancing Translational Sciences. She is an associate editor for *Toxicological Sciences* and *Toxicologic Pathology*, and she is editor-in-chief of the *ILAR Journal* (Institute for Laboratory Animal Research of the National Academy of Sciences). Dr. Davis attended Tuskegee University where she pursued a BS degree in chemistry and mathematics followed by a doctorate of veterinary medicine. She then received her Ph.D. in toxicology from the University of Illinois and obtained postdoctoral training in toxicologic pathology at the University of Maryland before starting her academic career.

James DeGregori, Ph.D., is a professor in the Department of Biochemistry and Molecular Genetics (faculty since 1997) and deputy director of the University of Colorado Cancer Center. He has degrees from the University of Texas at Austin (B.A., microbiology) and the Massachusetts Institute of

Technology (Ph.D., biology), and received postdoctoral training at Duke University. He holds the Courtenay and Lucy Patten Davis Endowed Chair in Lung Cancer Research. His lab studies the evolution of cancer, in the context of their adaptive oncogenesis model, with a focus on how aging, smoking, Down syndrome, and other insults influence cancer initiation and responses to therapy. In this model, mutations face fitness landscapes that vary with age or genetics, or following carcinogen exposure. These fitness landscapes are highly dependent on the state of the tissue microenvironment in which stem cells reside. The lab has developed this cancer model based on classic evolutionary principles, and has substantiated this model by theoretical, experimental, and computational studies. Additional studies in the lab seek to identify metabolic and signaling vulnerabilities in cancer, with a focus on acute myeloid leukemias, which can be exploited for the development of more effective therapies. For all of these studies, the lab leverages a variety of tools, including computational biology, genomics, metabolomics, cell biology, and biochemistry, leveraging both mouse models and human samples.

Nicole Deziel, Ph.D., M.H.S., is an associate professor in environmental health sciences at the Yale School of Public Health and a member of the Yale Cancer Center and the Yale Center for Perinatal, Pediatric and Environmental Epidemiology. Over the past 15 years, her research has involved applying existing and advanced statistical models, biomonitoring techniques, and environmental measurements to provide comprehensive and quantitative assessments of exposure to combinations of traditional and emerging environmental contaminants. Dr. Deziel's work involves the use of large administrative datasets in conjunction with detailed field-based studies. Her exposure assessment strategies aim to reduce exposure misclassification for epidemiologic studies, advancing understanding of relationships between exposure to environmental chemicals and risk of adverse health outcomes, particularly among women and children. She served as principal investigator of a study funded by the American Cancer Society evaluating co-exposures to multiple flame retardants, pesticides, and other persistent pollutants and thyroid cancer risk in adult women, and is now leading a project studying environmental exposures and pediatric thyroid cancers. She is also leading an interdisciplinary team of investigators on a project titled "Drinking water vulnerability and neonatal health outcomes in relation to oil and gas production in the Appalachian Basin," which is evaluating whether exposure to water contaminants from the process of hydraulic fracturing is associated with adverse human developmental and teratogenic effects. Dr. Deziel serves as associate editor for the *Journal of Exposure Science and Environmental Epidemiology* and is on the editorial board of *Environment International.* She is also a member of the National Academies of Sciences Standing Committee the Use of Emerging Science for Environmental Health Decisions.

Mark Dunn, M.B.A., is the executive vice president of the American Kennel Club (AKC) and is the managing director of AKC Reunite. Founded in 1884, the AKC is the oldest all-breed dog registry in the United States and the largest in the world. Mr. Dunn leads the AKC's efforts to meet the needs of breeders and dog owners. He also works with pet-industry leaders and international registry organizations to do good things for dogs and the people who love them, around the world. As part of those responsibilities, Mr. Dunn oversees AKC's DNA Program. The AKC has for more than 20 years harnessed the power of genotyping technology to ensure the integrity of its registry and to assist breeders with the accuracy of their breeding records. Mr. Dunn joined the AKC in 2009 as director of internal consulting. Previously he was director of engineering and quality at Qualex, a subsidiary of Eastman Kodak, and has over 20 years' experience leading operations, engineering, and business development teams.

Janice A. Dye, D.V.M., Ph.D., M.S., is a scientist within the EPA's Center for Public Health and Environmental Assessment. She is a board-certified veterinary internist whose clinical interests include comparative respiratory diseases, lung function testing, and airway cell biology as well as general internal medicine and infectious disease. Using animal model, animal sentinel, and in vitro cellular approaches, the purpose of her toxicological research is to increase understanding of mechanisms by which exposure to air pollutants, environmental agents, and nonenvironmental factors may contribute to increased susceptibility to developing adverse respiratory, cardiometabolic, or endocrine health outcomes.

Gary L. Ellison, Ph.D., M.P.H., is on detail to the National Institute of Environmental Health Sciences (NIEHS), where he has served as acting director of the Division of Extramural Research and Training since January 2021. His position of record is chief of the Environmental Epidemiology Branch (EEB) in the Epidemiology and Genomics Research Program, Division of Cancer Control and Population Sciences, at the National Cancer Institute. There, he oversees a program of extramural research focused on modifiable factors and risk of cancer. Dr. Ellison leads a group of program officers within EEB with expertise that spans all domains of the exposome, including the general external (e.g., broader social and policy context), specific external (e.g., lifestyle factors, environmental pollutants, chemical, physical, and infectious agents), and internal environments (e.g., microbiome, biomarkers of effect, early damage). Dr. Ellison has served as an ex-officio member of the National Advisory Environmental Health Sciences Council, a congressionally mandated body that advises the secretary of Health and Human Services

(HHS), director of the National Institutes of Health (NIH), and the director of the NIEHS on matters relating to research, research training, and career development supported by the NIEHS. He has received the NIH Director's Awards for the 2010 Gulf Oil Spill Response (2011); the NIH Working Group for the US-China Biomedical Research Cooperation Program (2013); and the GEOHealth Team for conceptualizing and implementing the Global Environmental and Occupational Health (GEOHealth) Program (2018). In 2014, he received an NIH Award of Merit for providing sustained leadership, scientific direction, and programmatic management for the Breast Cancer and the Environment research program.

William H. Farland, Ph.D., ATS, is an independent consultant in toxicology and environmental and public health, and a professor emeritus in environmental and radiological health sciences, School of Veterinary Medicine and Biomedical Sciences, Colorado State University (CSU). Formerly, Dr. Farland served as vice president for research at CSU from 2006–2013. Prior to that, he had a 27-year federal career at the Environmental Protection Agency (EPA), serving ultimately as the deputy assistant administrator for science in the Office of Research and Development, and acting agency science advisor in 2005. His tenure at the EPA was characterized by a commitment to the development of national and international approaches to research, testing, and assessment of the fate and effects of environmental agents. Dr. Farland holds a Ph.D. in cell biology and biochemistry from the University of California, Los Angeles. Throughout his career, he has served extensively on executive-level committees and advisory boards within the federal government, academia, and internationally. He is currently the chair of the Board on Environmental Studies and Toxicology for the National Academies of Sciences, Engineering, and Medicine in Washington, DC.

Caleb Finch, Ph.D., is ARCO Professor of Gerontology and Biological Sciences at the University of Southern California (USC), with adjunct appointments in the Departments of Anthropology, Molecular Biology, Neurobiology, Psychology, Physiology, and Neurology. His major research interest is the neurobiology of aging and human evolution. Dr. Finch received his undergraduate degree from Yale in 1961 (biophysics) and Ph.D. from Rockefeller University in 1969 (biology). His life's work is on the fundamental biology of human aging, starting in graduate school and continuing since 1972 at USC. His discoveries include oligomeric Abeta, a novel form of neurotoxicity of amyloid peptides in Alzheimer's disease; the role of shared inflammatory pathways in normal and pathological aging process; and the acceleration of aging processes by air pollution. Dr. Finch was founding director of the National Institute on Aging–funded USC Alzheimer's Disease Research Center (1984) and continues as co-principal investigator. He also cofounded Acumen Pharmaceuticals, which develops therapeutics for Alzheimer's disease. Fifteen of his mentored students hold senior positions in universities or pharmaceutical corporations. Dr. Finch has received most of the major awards in biomedical gerontology, including the Robert W. Kleemeier Award (1985), the Sandoz Premier Prize (1995), and the Irving Wright Award (1999). In 2018, the French Academy (EPHE) awarded him the doctorate *Honaris causis*. He has written six books, most recently *The Role of Global Air Pollution in Aging and Disease* (Academic Press, 2018). His current lab focus is on gene–environment interactions for brain aging, particularly air pollution components.

Marcia C. Haigis, Ph.D., is a professor in the Department of Cell Biology and the director of gender equity for faculty in science at Harvard Medical School. She obtained her Ph.D. in biochemistry from the University of Wisconsin and performed postdoctoral studies at the Massachusetts Institute of Technology studying mitochondrial metabolism. Dr. Haigis is an active member of the Dana Farber/Harvard Cancer Center, the Paul F. Glenn Center for the Biology of Aging Research, and the Ludwig Center at Harvard Medical School. Her research has made fundamental contributions to our understanding of how mitochondria mediate metabolic reprogramming in cancer, including identifying nodes of metabolic vulnerability in the control of fat oxidation in leukemia and metabolic recycling of ammonia to generate amino acids important for tumor growth. Most recently, her work has shed light on our understanding of how diet and environmental factors regulate anti-tumor immunity. She is the recipient of numerous honors and awards, including the Brookdale Leadership in Aging Award, the Ellison Medical Foundation New Scholar Award, the American Cancer Society Research Scholar Award, and the National Academy of Medicine Emerging Leaders in Health and Medicine Program.

Angela Hughes, D.V.M., Ph.D., serves as global science advocacy senior manager at Mars Petcare, where she focuses on educating people about the science behind the human–animal bond, as well as the development of new markers of health and disease in pets. She is a trained veterinary geneticist who pioneered the concept of genetically aligning potential breeding dogs to evaluate genetic diversity and launched this in a first-of-its-kind test called Optimal Selection™. Dr. Hughes completed her veterinary degree, veterinary genetics residency, Ph.D. in genetics, and held an associate clinical professor position at the University of California, Davis prior to joining Mars Petcare. She has been published in multiple academic publications including the *Journal of the American Veterinary Medical Association*, *PLOS Genetics*, and *PLOS ONE* and has contributed chapters for publication in *Veterinary Clinics of North America*

Small Animal Practice: Pediatrics and several editions of *Large Animal Internal Medicine*. Dr. Hughes's special interests include small animal and equine genetics and small animal reproduction and pediatrics.

Roy Jensen, M.D., was appointed director of The University of Kansas Cancer Center in 2004. As a result of a broad-based university, community, and regional effort, The University of Kansas Cancer Center was designated as a cancer center by the National Cancer Institute (NCI) in July 2012. Dr. Jensen is currently professor of pathology and laboratory medicine, professor of anatomy and cell biology, professor of cancer biology, and the William R. Jewell, MD Distinguished Kansas Masonic Professor, at the University of Kansas Medical Center. Prior to his appointment at Kansas, Jensen was a member of the Vanderbilt-Ingram Cancer Center and a faculty member in Pathology, Cell Biology, and Cancer Biology for 13 years. Dr. Jensen graduated from the Vanderbilt University School of Medicine in 1984 and remained there to complete a residency in anatomic pathology and a surgical pathology fellowship with Dr. David Page. Following his clinical training he accepted a postdoctoral fellowship at the National Cancer Institute in the laboratory of Dr. Stuart Aaronson. After joining the faculty at Vanderbilt University, Dr. Jensen's research interests focused on understanding the function of BRCA1 and BRCA2 and their role in breast neoplasia and in the characterization of premalignant breast disease at both the morphologic and molecular levels. He currently has more than 150 scientific publications and has lectured widely on the clinical and molecular aspects of breast cancer pathology. Dr. Jensen has served on numerous grant review panels, study sections, and site visit teams for the National Institutes of Health, the Department of Defense-Breast Cancer Research Program, the Medical Research Council of Canada, the California Breast Cancer Research Program, the Susan G. Komen Breast Cancer Foundation, and the Federation of American Societies for Experimental Biology. Jensen serves on the Science Policy and Governmental Affairs Committees for the American Association for Cancer Research (AACR) and is a member of the AACR Pathology Task Force and AACR Publications Committee. He served as a member of the Science Policy Working Group of the American Society for Investigative Pathology, and co-chaired the research committee for C-Change. In 2013, he was elected to the board of directors for the Association of American Cancer Institutes (AACI) and served as the president of AACI from 2018–2020. Dr. Jensen was chair of the NCI's Subcommittee A from 2018–2020 and also served on the Director's Working Group for the Board of Scientific Advisors to the National Cancer Institute. Finally, he is the chair of the University of Oklahoma Stephenson Cancer Center External Advisory Board.

Chad M. Johannes, D.V.M., DACVIM (SAIM, Oncology), is an associate professor of oncology at Iowa State University. His industry experience includes serving as a former medical director at Ariana Therapeutics, Inc. and coordination of the launch of Palladia®, the first FDA-approved veterinary cancer therapeutic, during his time with Pfizer Animal Health (now Zoetis). Dr. Johannes's practice experience includes primary care, specialty care, and academic settings. His areas of research interest include oncology therapeutic development, immunotherapeutics, and the effective management of treatment-related side effects.

Rena Jones, Ph.D., M.S., is an investigator in the Occupational and Environmental Epidemiology Branch, Division of Cancer Epidemiology & Genetics, at the National Cancer Institute (NCI), where her intramural research program seeks to identify and clarify the role of environmental exposures in the development of cancer. Dr. Jones's work relies on the application of geographic information systems and novel approaches to assess environmental exposures, a critical component of cancer epidemiology studies. She takes several approaches to improving long-term environmental exposure estimates, including optimizing the spatial accuracy of residential addresses and exposure sources; characterizing participant mobility and time spent in microenvironments; and incorporating information from surveys, regulatory environmental monitoring data, biomonitoring, and other secondary data sets. Her research program leads several large-scale, multidisciplinary efforts to characterize general population exposure to drinking-water contaminants and point source air pollution. In addition, she co-leads the NCI working groups focused on geospatial analyses and incorporation of new technologies for human exposure assessment in population studies. The novelty and quality of Dr. Jones's work has been recognized through multiple research awards, including the 2020 NCI Director's Intramural Innovation Award. She received her master's and doctoral degrees in epidemiology from the University at Albany, State University of New York.

Kurunthachalam Kannan, Ph.D., is a professor in the Department of Pediatrics, Division of Environmental Pediatrics, at New York University School of Medicine. He has published more than 780 research articles in peer-reviewed journals and 25 book chapters and has coedited a book. Dr. Kannan is the top five most highly cited researchers (ISI) in ecology/environment globally with an H-index of 135 (Google scholar) or 118 (Scopus). He is known for his work on the discovery of perfluorochemicals in the global environment, among several others. Currently his research is focused on biomonitoring of human exposure to organic pollutants. Dr. Kannan has won several medals for his stellar academic career, gold medals for his top rank in undergraduate academic career throughout, and to name a few, Governor's gold medal in 1986 and SETAC's Weston F. Roy Environmental Chemistry award in 1999, New York State Department of Health's Sturman Award for Excellence in Research in 2019. He has mentored more than 15 master's and doctoral level students and advised more than 60 postdoctoral research associates in his laboratory.

Norman Kleiman, Ph.D., M.S., works at the intersection of public health, radiation research, and ophthalmology, often using the eye as a model system to study the effects of environmental exposures, and radiation in particular, on human and animal health. For example, National Aeronautics and Space Administration– and Department of Energy–funded research projects were designed to better understand ocular risks, and radiation cataract in particular, underlying eye exposure to low doses of different kinds of radiation, e.g. X-rays and high-energy space radiation, (think, cosmic rays). Related human research in Dr. Kleiman's laboratory estimates relative risk of radiation cataract in medical professionals, such as interventional cardiologists and associated nursing personnel, following occupational exposure to X-rays during fluoroscopic imaging procedures. A collaborative study with Ukrainian colleagues examines ocular radiation risk in Chornobyl accident cleanup workers. Recently, new projects have examined health risks posed by exposure to radiation, heavy metals, and other environmental hazards in mice, voles, and semi-domesticated dogs living within the Chernobyl exclusion zone. In other areas related to eye pathology, a National Institute of Environmental Health Sciences (NIEHS)-funded project investigates the potential relationship between arsenic exposure and cataracts and recently reported significantly elevated arsenic concentrations in eye tissue. A recently funded NIEHS study examines the potentially carcinogenic heavy-metal risks associated with e-cigarette use. At a mechanistic level, Dr. Kleiman applies molecular and biochemical approaches to examine how environmental toxins, such as those from radiation, heavy metals, or e-cigarette use, cause DNA damage, misrepair, and mutagenesis and how individual genetic determinants influence risk. Overall, these investigations help in formulating appropriate risk policies and aid in development of human exposure guidelines as well as having important therapeutic implications for radio- and/or chemo-sensitive subsets of the human population. Among other responsibilities, Dr. Kleiman is a technical cooperation expert for the International Atomic Energy Agency and serves on scientific committees of the National Council on Radiation Protection and the International Commission on Radiological Protection.

Richard Lea, Ph.D., SFHEA, is currently a reader and associate professor in the School of Veterinary Medicine and Science at the University of Nottingham, UK, and is a professor of reproductive biology as of January 2022. Dr. Lea is chair of the School Committee for Animals and Research Ethics, deputy head of the Division of Global Health, and has been central to the development of the teaching curriculum in veterinary reproduction for over 15 years. Dr. Lea is also the chair of the Society for Reproduction and Fertility and actively promotes public awareness on environmental threats to reproductive health. Dr. Lea has over 30 years' experience in research into environmental influences on fertility and reproduction. His primary research program concerns the topical issue of environmental chemicals and their effects on mammalian reproductive well-being and his experimental approaches encompass both animal and human studies. Dr. Lea's primary research programs concern firstly the dog as a sentinel species for human exposure to household and industrial pollutants and secondly, the sheep as 'real-life' model for exposure to chemical mixtures in a commonly used agricultural fertilizer. Of note is the demonstration of a 26-year decline in dog semen quality that parallels that widely reported in the human, and maternal exposure linked perturbations in ovine female fetal reproductive development. These programs have been supported by grants awarded by the European Union and national UK charities. Currently, Dr. Lea is the Nottingham (UK) principal investigator on an R01 National Institutes of Health–funded study focused on multigenerational effects in sheep following maternal exposure to environmentally relevant chemical mixtures. Dr. Lea's complementary research paradigms suggest that the utilization of the sheep and dog in future research provides a means of investigating environmental influences on fertility in a manner complementary to essential human studies.

Amy K. LeBlanc, D.V.M., is a board-certified veterinary oncologist, and senior scientist and the director of the intramural National Cancer Institute's (NCI's) Comparative Oncology Program. In this position she conducts preclinical mouse and translational pet dog studies that are designed to inform the drug and imaging agent development path for human cancer patients, specifically those with osteosarcoma. She directly oversees the NCI Comparative Oncology Trials Consortium, which provides infrastructure necessary to connect participating veterinary academic institutions with stakeholders in drug development to execute fit-for-purpose comparative clinical trials in novel therapeutics and imaging agents. Her program provides support to several extramural NCI-funded initiatives including the Integrated Canine Data Commons and Cancer Moonshot–funded canine immunotherapeutic clinical trials conducted under the PRECINCT network.

Gary W. Miller, Ph.D., serves as vice dean for research strategy and innovation and professor of environmental health sciences at the Columbia University Mailman School of Public Health. He is an international leader on the exposome, the environmental analogue to the genome. Dr. Miller founded the first exposome center in the United States and wrote the first book on the topic. He has helped develop high-resolution mass spectrometry methods to provide an '-omic-scale analysis of the human exposome. He serves as codirector of Columbia's Irving Institute Precision Medicine Resource, which supports integration of environmental measures into clinical and translational research projects, and is a member of the National Institutes of Health All of Us Research Program Advisory Panel. Dr. Miller is the founding editor of the new journal *Exposome*, published by Oxford University Press.

Lisa Moses, VMD, DACVIM, is a veterinarian and animal-focused bioethicist. After nearly 30 years as a practicing veterinary specialist for the MSPCA Animal Medical Center in Boston, Dr. Moses became a faculty member at Harvard Medical School's Center for Bioethics. Dr. Moses is the chair of both the Animal Ethics Study Group at Yale's Interdisciplinary Center for Bioethics and the Harvard–Yale Animal Ethics Faculty Seminar, and she holds a visiting scientist appointment at The Broad Institute of MIT and Harvard. She completed a fellowship in bioethics at the Harvard Medical School Center for Bioethics and received her veterinary degree from the University of Pennsylvania. She also holds a faculty fellow position at Cummings Tufts School of Veterinary Medicine Center for Animals and Public Policy. Dr. Moses teaches and studies various aspects of veterinary medical and animal conservation ethics, most recently concentrating on research ethics where animals are both the subject and beneficiary of research investigations.

Elaine A. Ostrander, Ph.D., is the chief of the Cancer Genetics and Comparative Genomics Branch, and a distinguished senior investigator at the National Human Genome Research Institute of the National Institutes of Health. She has published more than 375 papers and won several awards, including a 2013 Genetics Society of America Medal, and was elected to the National Academy of Sciences in 2019. Her lab is interested in understanding the role that genomic variation plays in canine aging, morphology, behavior, and disease susceptibility. Its studies include evolution, genome architecture, breed formation, breed-specific disease, and the genetics of morphologic variation between breeds. Using genome sequencing, her lab shows that most breed-defining traits, such as body size, leg length, and so on, are controlled by small numbers of genes of large effect, and that most are also relevant for human health and biology. The lab's studies of breed-enriched diseases reveal the genetic underpinnings of disorders such as cancer, and have advanced studies of similar human disorders, while demonstrating the utility of the dog system for studies of human health. Finally, their collaborative studies of aging reveal conserved changes centering on developmental gene networks, which are sufficient to translate age and the effects of anti-aging interventions across multiple mammals. These studies establish methylation as a cross-species translator of the physiological milestones of aging.

Rodney Page, D.V.M., received his D.V.M. from Colorado State University and completed specialty training in the field of medical oncology in NYC. Dr. Page is board certified in internal medicine and oncology. He was a faculty member at North Carolina State University prior to his appointment at Cornell University as the founding director of The Sprecher Institute for Comparative Cancer Research. In 2005 Dr. Page was appointed chair of the Department of Clinical Sciences. He returned to Colorado as the director of the Flint Animal Cancer Center in 2010. Dr. Page's research interests have focused on a 'One Medicine' approach to cancer. He has served as principal investigator of the Golden Retriever Lifetime Study since 2008 and has led national efforts to bring translational and comparative oncology to a greater audience. He is the 2019 recipient of the AVMA/AKC Career Achievement Award in Canine Research.

Daniel Promislow, D.Phil., is a professor in the Department of Biology and in the Department of Laboratory Medicine & Pathology at the University of Washington. Since receiving his D.Phil. in 1990 at the University of Oxford, he has focused on the study of aging. He began his career on faculty in the Department of Genetics at the University of Georgia from 1995–2013, then he moved to the University of Washington. His research uses evolutionary genetics and systems biology approaches to understand how genes and environment shape aging and age-related disease in natural populations. In addition to his lab-based research on *Drosophila*, Dr. Promislow is principal investigator and codirector of the Dog Aging Project, an National Institutes of Health/National Institute on Aging U19-funded nationwide research program to understand the determinants of healthy aging in tens of thousands of companion dogs.

Peter Rabinowitz, M.D., M.P.H., is a physician and professor in the University of Washington Schools of Public Health and Medicine. He came to UW 8 years ago to found the UW Center for One Health Research. The center conducts research and training to explore 'One Health' connections between the health of humans, animals, and the environments we share with other species. A key mission of the center is to find new ways that humans and animals can safely and sustainably coexist in a changing environment.

Audrey Ruple, D.V.M., M.S., Ph.D., DipACVPM, MRCVS, is an associate professor in the Department of Population Health Sciences in the Virginia–Maryland College of Veterinary Medicine at Virginia Tech. Her research focus is in the area of 'One Health,' the intersection of human, animal, and environmental health. She has a particular interest in comparative biomedical aspects of cancer and aging and she uses companion dogs as a model system to better understand why cancers occur and how we can all—humans and animals— age better. Dr. Ruple is a licensed, clinical veterinarian and is a diplomate of the American College of Veterinary Preventive Medicine. She obtained her D.V.M., M.S., and Ph.D. degrees from Colorado State University and is a member of the Royal College of Veterinary Surgeons in the United Kingdom.

Elizabeth Ryan, Ph.D., received her Ph.D. in toxicology from the University of Rochester School of Medicine and is an associate professor in the Colorado State University College of Veterinary Medicine and Biomedical Sciences. She leads a multidisciplinary lab team that studies environmental exposures, including those from foods for impacts on gut microbiota, mucosal immunity, and for protection against infectious and chronic diseases. She conducts cancer research in laboratory models, companion animals, and people in connection with cancer control and prevention initiatives at the University of Colorado Cancer Center. Her team implements dietary interventions with rice bran and legumes (e.g., beans, cowpeas) to understand the impacts of these nutrient-dense, phytochemical and fiber–rich foods on gut microbiome metabolism. The lab utilizes cutting-edge technologies such as metabolomics to evaluate a suite of microbial and chemical exposures from the diet and environment. Her research on native gut probiotic metabolism of foods and gut associated microbiota in response to dietary interventions across the life span is currently supported by the National Institutes of Health, the National Institute of Food and Agriculture, and the Thrasher Fund.

Norman E. "Ned" Sharpless, M.D., was officially sworn in as the 15th director of the National Cancer Institute (NCI) on October 17, 2017. Prior to his appointment, Dr. Sharpless served as the director of the Lineberger Comprehensive Cancer Center at the University of North Carolina (UNC). Dr. Sharpless was a Morehead Scholar at UNC–Chapel Hill and received his undergraduate degree in mathematics. He went on to pursue his medical degree from the UNC School of Medicine, graduating with honors and distinction in 1993. He then completed his internal medicine residency at the Massachusetts General Hospital and a hematology/oncology fellowship at Dana-Farber/Partners Cancer Care, both of Harvard Medical School in Boston. After 2 years on the faculty at Harvard Medical School, he joined the faculty of the UNC School of Medicine in the Departments of Medicine and Genetics in 2002. He became the Wellcome Professor of Cancer Research at UNC in 2012. Dr. Sharpless is a member of the Association of American Physicians and the American Society for Clinical Investigation, and is a fellow of the Academy of the American Association of Cancer Research. He has authored more than 160 original scientific papers, reviews, and book chapters, and is an inventor on 10 patents. He cofounded two clinical-stage biotechnology companies: G1 Therapeutics and Sapere Bio (formerly HealthSpan Diagnostics). He served as acting commissioner for Food and Drugs at the U.S. Food and Drug Administration for 7 months in 2019, before returning to the NCI directorship.

Wendy C. Shelton, D.V.M., M.P.H., brings experience in clinical medicine, medical devices, drug development, business development, public health, government policy, and project management. She provides strategic expertise regarding the interrelationships of these sectors. Dr. Shelton began her professional life as a practicing veterinarian after graduating from the University of California at Davis School of Veterinary Medicine in 1981. She was a small-animal practitioner and small-business owner for more than 12 years. In 1993, Dr. Shelton accepted a position on the Board of Directors of Integrated Surgical Systems, developers and manufacturers of the world's first computer-guided surgical robot, ROBODOC®. She stayed with the company in numerous capacities (VP, research and development; VP, medical affairs; acting CEO) until it went public. There she gained experience preparing applications for FDA approval, conducting animal and human clinical trials, creating an iso9000 manufacturing facility, and developing a European market for the device. The company was the recipient of the Computerworld-Smithsonian Award for Excellence in Medical IT, and the device was collected by the museum. Dr. Shelton subsequently spent several years combining part-time equine practice and new therapeutic product development, and then earned her master of public health degree from the UC Davis School of Medicine. A brief position at the California Department of Health Services, Department of Infectious Diseases, Office of the Public Health Veterinarian, working on West Nile virus surveillance systems followed. Dr. Shelton was a fully funded Congressional Fellow, sponsored by the American Veterinary Medical Association and placed by the American Association for the Advancement of Science in the office of Senator Joseph Lieberman of Connecticut in 2004–2005. While in the senator's office, she participated in the genesis of Senate Bill 975, or BioShield II, the massive legislative initiative designed to create a countermeasures industry to address both bioterror and naturally occurring public health threats, and was primary author of several titles. From Capitol Hill, Dr. Shelton was recruited to Fabiani & Company, a DC lobbying firm, where she helped build a practice that matched growing life sciences companies with government funding sources. She advised dozens of health care, drug, device, and product companies and academic institutions regarding government and business relations and helped secure over $250 million in grants and contracts more than 5 years. More recently, Dr. Shelton worked in Silicon Valley developing veterinary applications for in silico biosimulation models. She served as vice president of corporate communications, government relations, and veterinary applications at Entelos Holding Corporation in San Mateo for more than 2 years—engaging Mars Petcare to create a virtual dog. Dr. Shelton is the founding principal of Virtual Beast Consulting (VBC) based in Truckee, California. Areas of focus include promotion of the study of companion animals as research models to improve understanding of human and animal diseases and treatments—the embodiment of the One Medicine/One Health concept. An example of this is the National Cancer Policy Forum's 2015 workshop titled The Role of Clinical Studies for Pets with Naturally Occurring Tumors in Translational Cancer Research, initiated by VBC's principal on behalf of the Flint Animal Cancer Center at Colorado State University (CSU). Dr. Shelton continues to consult with CSU, providing strategic support for Comparative Oncology and One Health, actively representing Flint in the recently formed CORC—the Comparative Oncology Research Consortium that pairs veterinary schools and National Cancer Institute–designated cancer centers for research funding. Dr. Shelton also represents CSU with the CTSA One Health Alliance, serving on the advocacy subcommittee.

Heather M. Stapleton, Ph.D., is an environmental chemist and exposure scientist in the Nicholas School of the Environment at Duke University. Her research interests focus on identification of halogenated and organophosphate chemicals in consumer products and building materials and estimation of human exposure, particularly in vulnerable populations such as pregnant women and children. Her laboratory specializes in analysis of environmental and biological tissues for organic contaminants to support environmental health research. Her research projects seek to understand how chronic exposure to chemical mixtures impact human health, with an emphasis on elucidating effects on thyroid hormone dysregulation and associations with thyroid disease. She received an early career award from the National Institute of Environmental Health Sciences in 2008, called the Outstanding New Environmental Scientist award, which helped to propel her research career. In 2012 she testified in front of the U.S. Senate Environment and Public Works Committee on human exposure and toxicity on new-use flame retardant chemicals used in commerce and in 2014 she helped to develop a resource for the general public to support free testing for flame retardant chemicals in consumer products. She currently serves as the director for the Duke Superfund Research Center, and as director of the Duke Environmental Analysis Laboratory.

Anne Thessen, Ph.D., is a visiting associate professor at the University of Colorado Anschutz Medical Campus. She received her Ph.D. in oceanography and shifted toward data science while working for the Encyclopedia of Life and the Census of Marine Life. Later, she started her own data science consulting company and operated that for 5 years before joining the Translational and Integrative Sciences Lab under Dr. Melissa Haendel.

Frank A. von Hippel, Ph.D., is a professor of environmental health sciences in the Mel & Enid Zuckerman College of Public Health and the lead of the One Health Research Initiative at the University of Arizona. Dr. von Hippel was born and raised in Alaska, received his A.B. in biology at Dartmouth College in 1989, and his Ph.D. in integrative biology at the University of California, Berkeley in 1996. He taught for Columbia University (1996–1999), the University of Alaska Anchorage (2000–2016), and Northern Arizona University (2016–2021) before moving to the University of Arizona in 2021. Dr. von Hippel has taught ecology field courses in over 20 countries, and has conducted research in the Americas, Africa, and Australia. He conducts research at the nexus of ecotoxicology, mechanisms of toxicity, and health disparities, with a focus on indigenous and underserved communities. Dr. von Hippel is the author of *The Chemical Age* (University of Chicago Press, 2020; https://frankvonhippel.github.io/pubs.html) and he is the creator and host of the Science History Podcast (https://podcasts.apple.com/us/podcast/science-history-podcast/id1325288920).

Joseph Wakshlag, Ph.D., M.S., started his academic career receiving a B.S. and M.S. from Montclair State University. He then attended the Cornell College of Veterinary Medicine, graduating in 1998. He continued his residency training in both pathology and nutrition, as well as receiving his Ph.D. in pharmacology in 2005. He became a diplomate in the College of Veterinary Nutrition in 2008 and furthered his board certification as a diplomate in the College of Veterinary Sports Medicine and Rehabilitation in 2010, and is currently a professor at Cornell University College of Veterinary Medicine. He has been teaching both basic veterinary nutrition and small animal clinical nutrition at the Cornell College of Veterinary Medicine for nearly 20 years since his residency in Small Animal Clinical Nutrition. He is the current service chief for clinical nutrition at the college and also does service work for the Sports Medicine and Rehabilitation Service at the Cornell University Hospital for Animals. His background in sports medicine and nutrition has produced many publications on working dogs, obesity, canine cancer cell biology, the canine GI microbiome, and arthritis management.

Cheryl Lyn Walker, Ph.D., holds the Alkek Presidential Chair in Environmental Health and is the founder and director of the Center for Precision Environmental Health at the Baylor College of Medicine in Houston, Texas.

She also directs the National Institute of Environmental Health Sciences P30 Gulf Coast Center for Precision Environmental Health (https://gc-cpeh.org). Dr. Walker has over 200 publications in the scientific literature and is an elected member of the National Academy of Medicine. Her research on gene–environment interactions and environmental epigenomics has led to new insights into how early life exposures reprogram the developing epigenome to alter disease susceptibility across the life course. She has been recognized with the Roy O. Greep Laureate Award from the Endocrine Society, the Leading Edge in Basic Science Award from the Society of Toxicology (SOT), and the Distinguished Scientist Award from the American College of Toxicology. In addition to her research accomplishments, she has held significant professional administrative and leadership positions including president of SOT, president of Women in Cancer Research for the American Association for Cancer Research, and as the founding chair of the Systemic Injury from Environmental Exposures Study Section for the Center for Scientific Review of the National Institutes of Health. Dr. Walker has also served on the boards of Scientific Advisors and Scientific Councilors of the National Cancer Institute and National Toxicology Program.